Sie Chin Tjong
Carbon Nanotube Reinforced Composites

Related Titles

Hierold, C. (ed.)

Carbon Nanotube Devices

Properties, Modeling, Integration and Applications

2008
ISBN: 978-3-527-31720-2

Haley, M. M., Tykwinski, R. R. (eds.)

Carbon-Rich Compounds

From Molecules to Materials

2006
ISBN: 978-3-527-31224-5

Reich, S., Thomsen, C., Maultzsch, J.

Carbon Nanotubes

Basic Concepts and Physical Properties

2004
ISBN: 978-3-527-40386-8

Roth, S., Carroll, D.

One-Dimensional Metals

Conjugated Polymers, Organic Crystals, Carbon Nanotubes

2004
ISBN: 978-3-527-30749-4

Sie Chin Tjong

Carbon Nanotube Reinforced Composites

Metal and Ceramic Matrices

WILEY-VCH Verlag GmbH & Co. KGaA

The Author

Prof. Sie Chin Tjong
City University of Hong Kong
Department of Physics and Materials Science
Hongkong, PR China

All books published by **Wiley-VCH** are carefully produced. Nevertheless, authors, editors, and publisher do not warrant the information contained in these books, including this book, to be free of errors. Readers are advised to keep in mind that statements, data, illustrations, procedural details or other items may inadvertently be inaccurate.

Library of Congress Card No.: applied for

British Library Cataloguing-in-Publication Data
A catalogue record for this book is available from the British Library.

Bibliographic information published by the Deutsche Nationalbibliothek
Die Deutsche Nationalbibliothek lists this publication in the Deutsche Nationalbibliografie; detailed bibliographic data are available on the Internet at http://dnb.d-nb.de.

© 2009 WILEY-VCH Verlag GmbH & Co. KGaA, Weinheim

All rights reserved (including those of translation into other languages). No part of this book may be reproduced in any form – by photoprinting, microfilm, or any other means – nor transmitted or translated into a machine language without written permission from the publishers. Registered names, trademarks, etc. used in this book, even when not specifically marked as such, are not to be considered unprotected by law.

Cover Design Spieszdesign, Neu-Ulm, Germany
Typesetting Thomson Digital, Noida, India
Printing Strauss GmbH, Mörlenbach
Binding Litges & Dopf GmbH, Heppenheim

Printed in the Federal Republic of Germany
Printed on acid-free paper

ISBN: 978-3-527-40892-4

Contents

Preface *IX*
List of Abbreviations *XI*

1	**Introduction** *1*	
1.1	Background *1*	
1.2	Types of Carbon Nanotubes *2*	
1.3	Synthesis of Carbon Nanotubes *5*	
1.3.1	Electric Arc Discharge *5*	
1.3.2	Laser Ablation *7*	
1.3.3	Chemical Vapor Deposition *9*	
1.3.3.1	Thermal CVD *11*	
1.3.3.2	Plasma-enhanced CVD *14*	
1.3.3.3	Laser assisted CVD *14*	
1.3.3.4	Vapor Phase Growth *14*	
1.3.3.5	Carbon Monoxide Disproportionation *16*	
1.3.4	Patent Processes *17*	
1.4	Purification of Carbon Nanotubes *18*	
1.4.1	Purification Processes *18*	
1.4.2	Materials Characterization *20*	
1.5	Mechanical Properties of Carbon Nanotubes *25*	
1.5.1	Theoretical Modeling *25*	
1.5.2	Direct Measurement *26*	
1.6	Physical Properties of Carbon Nanotubes *29*	
1.6.1	Thermal Conductivity *29*	
1.6.2	Electrical Behavior *30*	
1.7	Potential and Current Challenges *32*	
	References *33*	
2	**Carbon Nanotube–Metal Nanocomposites** *43*	
2.1	Overview *43*	
2.2	Importance of Metal-Matrix Nanocomposites *45*	

Carbon Nanotube Reinforced Composites: Metal and Ceramic Matrices. Sie Chin Tjong
Copyright © 2009 WILEY-VCH Verlag GmbH & Co. KGaA, Weinheim
ISBN: 978-3-527-40892-4

2.3	Preparation of Metal-CNT Nanocomposites	46
2.4	Aluminum-Based Nanocomposites	47
2.4.1	Spray Forming	47
2.4.2	Powder Metallurgy Processing	51
2.4.3	Controlled Growth of Nanocomposites	57
2.4.4	Severe Plastic Deformation	57
2.5	Magnesium-Based Nanocomposites	61
2.5.1	The Liquid Metallurgy Route	61
2.5.1.1	Compocasting	61
2.5.1.2	Disintegrated Melt Deposition	61
2.5.2	Powder Metallurgy Processing	62
2.5.3	Friction Stir Processing	64
2.6	Titanium-Based Nanocomposites	64
2.7	Copper-Based Nanocomposites	65
2.7.1	Liquid Infiltration	66
2.7.2	Mechanical Alloying	66
2.7.3	Molecular Level Mixing	68
2.7.4	Electrodeposition	71
2.7.5	Patent Process	72
2.8	Transition Metal-Based Nanocomposites	73
2.8.1	Ni-Based Nanocomposites	73
2.8.2	Co-Based Nanocomposites	76
2.8.3	Fe-Based Nanocomposites	77
	References	80
3	**Physical Properties of Carbon Nanotube–Metal Nanocomposites**	**89**
3.1	Background	89
3.1.1	Thermal Response of Metal-Matrix Microcomposites	91
3.2	Thermal Behavior of Metal-CNT Nanocomposites	93
3.2.1	Aluminum-Based Nanocomposites	93
3.2.2	Tin-Based Nanosolder	95
3.3	Electrical Behavior of Metal-CNT Nanocomposites	98
	References	100
4	**Mechanical Characteristics of Carbon Nanotube–Metal Nanocomposites**	**103**
4.1	Strengthening Mechanism	103
4.2	Tensile Deformation Behavior	106
4.2.1	Aluminum-Based Nanocomposites	106
4.2.2	Magnesium-Based Nanocomposites	110
4.2.3	Copper-Based Nanocomposites	112
4.2.4	Nickel-Based Nanocomposites	116
4.3	Comparison with Nanoparticle-Reinforced Metals	117
4.4	Wear	119
	References	127

5	**Carbon Nanotube–Ceramic Nanocomposites** *131*
5.1	Overview *131*
5.2	Importance of Ceramic-Matrix Nanocomposites *133*
5.3	Preparation of Ceramic-CNT Nanocomposites *136*
5.4	Oxide-Based Nanocomposites *138*
5.4.1	Alumina Matrix *138*
5.4.1.1	Hot Pressing/Extrusion *138*
5.4.1.2	Spark Plasma Sintering *142*
5.4.1.3	Plasma Spraying *147*
5.4.1.4	Template Synthesis *149*
5.4.2	Silica Matrix *150*
5.4.3	Titania Matrix *155*
5.4.4	Zirconia Matrix *156*
5.5	Carbide-Based Nanocomposites *157*
5.5.1	Silicon Carbide Matrix *157*
5.6	Nitride-Based Nanocomposites *160*
5.6.1	Silicon Nitride Matrix *160*
	References *161*

6	**Physical Properties of Carbon Nanotube–Ceramic Nanocomposites** *169*
6.1	Background *169*
6.2	Electrical Behavior *171*
6.3	Percolation Concentration *172*
6.4	Electromagnetic Interference Shielding *177*
6.5	Thermal Behavior *179*
	References *182*

7	**Mechanical Properties of Carbon Nanotube–Ceramic Nanocomposites** *185*
7.1	Fracture Toughness *185*
7.2	Toughening and Strengthening Mechanisms *187*
7.3	Oxide-Based Nanocomposites *189*
7.3.1	Alumina Matrix *189*
7.3.1.1	Deformation Behavior *189*
7.3.2	Silica Matrix *200*
7.4	Carbide-Based Nanocomposites *201*
7.5	Nitride-Based Nanocomposites *205*
7.6	Wear Behavior *207*
	References *212*

8	**Conclusions** *215*
8.1	Future Prospects *215*
8.2	Potential Applications of CNT–Ceramic Nanocomposites *217*

8.2.1	Hydroxyapatite–CNT Nanocomposites 218
8.3	Potential Applications of CNT–Metal Nanocomposites 223
	References 224

Index 227

Preface

Carbon nanotubes are nanostructured carbon materials having large aspect ratios, extremely high Young's modulus and mechanical strength, as well as superior electrical and thermal conductivities. Incorporation of a small amount of carbon nanotube into metals and ceramics leads to the formation of high performance and functional nanocomposites with enhanced mechanical and physical properties. Considerable attention has been applied to the development and synthesis of carbon nanotube-reinforced composites in the past decade. However, there is no published book that deals exclusively with the fundamental issues and properties of carbon nanotube-reinforced metals and ceramics. This book mainly focuses on the state-of-the-art synthesis, microstructural characterization, physical and mechanical properties and application of carbon nanotube-reinforced composites. The various synthetic and fabrication techniques, dispersion of carbon nanotubes in composite matrices, morphological and interfacial behaviors are discussed in detail. Manufacturing of these nanocomposites for commercial applications is still in an embryonic stage. Successful commercialization of such nanocomposites for industrial and clinical applications requires a better understanding of the fundamental aspects. With a better understanding of the processing–structure–property relationship, carbon nanotube-reinforced composites with predicted and tailored physical/mechanical properties as well as good biocompatibility can be designed and fabricated. This book serves as a valuable and useful reference source for chemists, materials scientists, physicists, chemical engineers, electronic engineers, mechanical engineers and medical technologists engaged in the research and development of carbon nanotube-reinforced metals and ceramics.

2008
City University of Hong Kong

S. C. Tjong,
CEng CSci FIMMM

Carbon Nanotube Reinforced Composites: Metal and Ceramic Matrices. Sie Chin Tjong
Copyright © 2009 WILEY-VCH Verlag GmbH & Co. KGaA, Weinheim
ISBN: 978-3-527-40892-4

List of Abbreviations

AAO	Anodic aluminum oxide
AFM	atomic force microscopy
ASTM	American Society for Testing and Materials
CB	carbon black
CF	carbon fiber
CMC	ceramic-matrix composite
CNB	chevron notched beam
CNF	carbon nanofiber
CNT	Carbon nanotube
$C_{16}TAB$	hexacetyltrimethyl ammonium bromide
CTE	coefficient of thermal expansion
CVD	chemical vapor deposition
d.c.	direct current
DMD	disintegrated melt deposition
DMF	N-dimethylformaldehyde
DSC	differential scanning calorimetry
DTG	differential thermogravimetry
ECAP	equal channel angle pressing
ECR-MW	electron cyclotron resonance microwave
EDS	energy-dispersive spectroscopy
EDX	energy-dispersive X-ray
EMI	Electromagnetic interference
EPD	electrophoretic deposition
f-SWNT	funtionalized SWNT
FED	field emission displays
FM	fully melted
FSP	Friction stir processing
FTIR	Fourier-transform infrared
FWHM	full-width at half-maximum
GF	graphite flake
GPTMS	(3-glycidoxypropyl)trimethoxysilane
HA	hydroxyapatite
HCP	hexagonal close-packed

Carbon Nanotube Reinforced Composites: Metal and Ceramic Matrices. Sie Chin Tjong
Copyright © 2009 WILEY-VCH Verlag GmbH & Co. KGaA, Weinheim
ISBN: 978-3-527-40892-4

HIP	hot isostatic pressing
HPC	hydroxypropylcellulose
HPT	high pressure torsion
HVOF	high velocity oxyfuel spraying
IC	integrated circuit
LCVD	Laser assisted CVD
LEFM	Linear elastic fracture mechanics
MA	mechanical alloying
MEMS	micro-electro-mechanical system
MMC	Metal-matrix composite
MWNT	Multi-walled carbon nanotubes
PAA	poly(acrylic acid)
PAN	polyacrylonitrile
PCA	process control agent
PCB	printed circuit board
PCS	polycarbosilane
PDC	polymer derived ceramic
PECVD	Plasma-enhanced CVD
PEI	polyethyleneamine
PIP	polymer infiltration and pyrolysis
PLV	pulsed laser vaporization
PM	partially melted
PM	powder metallurgy
PSF	plasma spray forming
PSO	polysiloxane
PSZ	polysilazane
PVD	physical vapor deposition
PVP-4	poly(4-vinylpyridine)
r.d.	relative density
r.f.	radio frequency
RBM	radial breathing mode
ROM	rule of mixture
RPS	reinforcement particle size
SDS	sodium dodecyl sulfate
SE	shielding effectiveness
SENB	single edge notched
SEPB	single edge precracked beam
SEVNB	single edge V-notched beam
SPS	Spark plasma sintering
TBC	thermal barrier coating
TCP	tricalcium phosphate
TEM	Transmission electron microscopy
TEOS	silicon tetraethyl orthosilicate
TGA	thermogravimetrical analysis
TIM	thermal interface material

TTCP	tetracalcium phosphate
UV	ultraviolet
UV-vis-NIR	Ultraviolet-visible-near infrared
VGCF	vapor grown carbon nanofiber
VIF	Vickers indentation fracture
VLS	vapor–liquid–solid
XRD	X-ray diffraction

1
Introduction

1.1
Background

Composite materials with a ceramic or metal matrix offer significant performance advantages over monolithic ceramics or metals. Structural ceramics exhibit high mechanical strength, superior temperature stability and good chemical durability. However, ceramics have low fracture toughness because of their ionic or covalent bonding. The intrinsic brittleness of ceramics limits their applications in industrial sectors. In this context, discontinuous fibers or whiskers are incorporated into ceramics in order to reinforce and toughen them. Upon application of external stress, matrix microcracking occurs followed by debonding of fibers from the matrix, and subsequent bridging of fibers across matrix cracks. These processes contribute to major energy dissipating events, thereby improving the fracture toughness of ceramics. Recently, metal-matrix composites (MMCs) have become increasingly used for applications in the automotive and aerospace industries because of their high specific modulus, strength and thermal stability. MMCs are reinforced with relatively large volume fractions of continuous fiber, discontinuous fibers, whiskers or particulates. The incorporation of ceramic reinforcement into the metal matrix generally leads to enhancement of strength and stiffness at the expense of fracture toughness. Further enhancement in mechanical strength of composites can be achieved by using nanostructured ceramic particles [1–4].

With tougher environmental regulations and increasing fuel costs, weight reduction in composites has become an important issue in the design of composite materials. The need for advanced composite materials having enhanced functional properties and performance characteristics is ever increasing in industrial sectors. Since their discovery by Ijima in 1991 [5], carbon nanotubes (CNTs) with high aspect ratio, large surface area, low density as well as excellent mechanical, electrical and thermal properties have attracted scientific and technological interests globally. These properties have inspired interest in using CNTs as reinforcing materials for polymer-, metal- or ceramic-matrix composites to obtain light-weight structural materials with enhanced mechanical, electrical and thermal properties [6–8]. Composite materials with at least one of their constituent phases being less than

Carbon Nanotube Reinforced Composites: Metal and Ceramic Matrices. Sie Chin Tjong
Copyright © 2009 WILEY-VCH Verlag GmbH & Co. KGaA, Weinheim
ISBN: 978-3-527-40892-4

100 nm are commonly termed "nanocomposites". Remarkable improvements in the mechanical and physical properties of polymer-, metal- and ceramic nanocomposites can be achieved by adding very low loading levels of nanotubes. So far, extensive studies have been conducted on the synthesis, structure and property of CNT-reinforced polymers. The effects of CNT additions on the structure and property of metals and ceramics have received increasing attention recently.

1.2 Types of Carbon Nanotubes

Hybridization of the carbon atomic orbital in the forms of sp, sp^2 and sp^3 produces different structural forms or allotropes [9] (Figure 1.1). The sp-hybridization (carbyne) corresponds to a linear chain-like arrangement of atomic orbital. Carbon in the form of diamond exhibits a sp^3-type tetrahedral covalent bonding. Each carbon atom is linked to four others at the corners of a tetrahedron via covalent bonding. This structure accounts for the extremely high hardness and density of diamond. The bonding in graphite is sp^2, with each atom joined to three neighbors in a trigonal

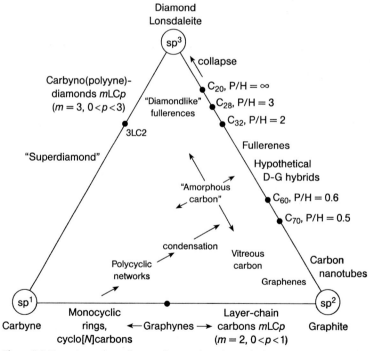

Figure 1.1 Tentative carbon allotropy diagram based on valence bond hybridization. P/H corresponds to the ratio of pentagonal/hexagonal rings. Reproduced with permission from [9]. Copyright © (1997) Elsevier.

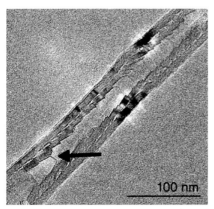

Figure 1.2 Transmission electron micrograph of the double-layer carbon nanofiber having a truncated cone structure (indicated by an arrow). Reproduced with permission from [13]. Copyright © (2006) Springer Verlag.

planar arrangement to form sheets of hexagonal rings. Individual sheets are bonded to one another by weak van der Waals forces. As a result, graphite is soft, and displays electrical conductive and lubricating characteristics. All other carbon forms are classified into intermediate or transitional forms in which the degree of hybridization of carbon atoms can be expressed as sp^n ($1 < n < 3$, $n \neq 2$). These include fullerenes, carbon onions and carbon nanotubes [9]. Fullerene is made up of 60 carbon atoms arranged in a spherical net with 20 hexagonal faces and 12 pentagonal faces, forming a truncated icosahedral structure [10, 11].

Carbon nanotubes are formed by rolling graphene sheets of hexagonal carbon rings into hollow cylinders. Single-walled carbon nanotubes (SWNT) are composed of a single graphene cylinder with a diameter in the range of 0.4–3 nm and capped at both ends by a hemisphere of fullerene. The length of nanotubes is in the range of several hundred micrometers to millimeters. These characteristics make the nanotubes exhibit very large aspect ratios. The strong van der Waals attractions that exist between the surfaces of SWNTs allow them to assemble into "ropes" in most cases. Nanotube ropes may have a diameter of 10–20 nm and a length of 100 μm or above. Multi-walled carbon nanotubes (MWNT) comprise 2 to 50 coaxial cylinders with an interlayer spacing of 0.34 nm The diameter of MWNTs generally ranges from 4 to 30 nm [12]. The arrangement of concentric graphene cylinders in MWNTs is somewhat similar to that of Russian doll. In contrast, a nanofiber consists of stacked curved graphite layers that form cones or cups (Figure 1.2). The stacked cone and cup structures are commonly referred to as "herringbone" and "bamboo" nanofibers [13, 14].

Conceptually, the graphene sheets can be rolled into different structures, that is, zig-zag, armchair and chiral. Accordingly, the nanotube structure can be described by a chiral vector ($\vec{C}_h$) defined by the following equation:

$$\vec{C}_h = n\vec{a}_1 + m\vec{a}_2 \tag{1.1}$$

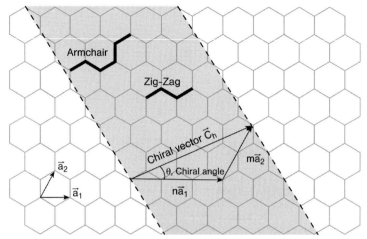

Figure 1.3 Schematic diagram showing chiral vector and chiral angle in a rolled graphite sheet with a periodic hexagonal structure. Reproduced with permission from [15]. Copyright © (2001) Elsevier.

where $\vec{a}_1$ and $\vec{a}_2$ are unit vectors in a two-dimensional hexagonal lattice, and n and m are integers. Thus, the structure of any nanotube can be expressed by the two integers n, m and chiral angle, θ (Figure 1.3). When $n = m$ and $\theta = 30°$, an armchair structure is produced. Zig-zag nanotubes can be formed when m or $n = 0$ and $\theta = 0°$ while chiral nanotubes are formed for any other values of n and m, having θ between $0°$ and $30°$ [15]. Mathematically, the nanotube diameter can be written as [16]:

$$d = \frac{a\sqrt{m^2 + n^2 + nm}}{\pi} \tag{1.2}$$

where a is the lattice constant in the graphene sheet, and $a = \sqrt{3}a_{C-C}$; a_{C-C} is the carbon–carbon distance (1.421 Å). The chiral angle, θ, is given by:

$$\tan\theta = \frac{\sqrt{3}m}{2n + m} \tag{1.3}$$

The electrical properties of CNTs vary from metallic to semiconducting, depending on the chirality and diameter of the nanotubes.

The demand for inexpensive carbon-based reinforcement materials is raising new challenges for materials scientists, chemists and physicists. In the past decade, vapor grown carbon nanofibers (VGCFs) with diameters ranging from 50 to 200 nm have been synthesized [17]. They are less crystalline with a stacked cone or cup structure, while maintaining acceptable mechanical and physical properties. Compared with SWNTs and MWNTs, VGCFs are available at a much lower cost because they can be mass-produced catalytically using gaseous hydrocarbons under relatively controlled conditions. VGCFs have large potential to override the cost barrier that has prevented widespread application of CNTs as reinforcing materials in industries.

1.3
Synthesis of Carbon Nanotubes

1.3.1
Electric Arc Discharge

Vaporization of carbon from solid graphite Sources into a gas phase can be achieved by using electric arc discharge, laser and solar energy. Solar energy is rarely used as the vaporizing Source for graphite, because it requires the use of a tailor-made furnace to concentrate an intense solar beam for vaporizing graphite target and metal catalysts in an inert atmosphere [18]. In contrast, the electric arc discharge technique is the simplest and less expensive method for fabricating CNTs. In the process, an electric (d.c.) arc is formed between two high purity graphite electrodes under the application of a larger current in an inert atmosphere (helium or argon). The high temperature generated by the arc causes vaporization of carbon atoms from anode into a plasma. The carbon vapor then condenses and deposits on the cathode to form a cylinder with a hard outer shell consisting of fused material and a softer fibrous core containing nanotubes and other carbon nanoparticles [19]. The high reaction temperature promotes formation of CNTs with a higher degree of crystallinity.

The growth mechanism of catalyst-free MWNTs is not exactly known. The nucleation stage may include the formation of C_2 precursor and its subsequent incorporation into the primary graphene structure. On the basis of transmission electron microscopy (TEM) observations, Ijiima and coworkers proposed the "open-end" growth mechanism in which carbon atoms are added at the open ends of the tubes and the growing ends remains open during growth. The thickening of the tube occurs by the island growth of graphite basal planes on existing tube surfaces. Tube growth terminates when the conditions are unsuitable for the growth [20, 21]. Generally, the quality and yield of nanotubes depend on the processing conditions employed, such as efficient cooling of the cathode, the gap between electrodes, reaction chamber pressure, uniformity of the plasma arc, plasma temperature, and so on [22].

Conventional electric arc discharge generally generates unstable plasma because it induces an inhomogeneity of electric field distribution and a discontinuity of the current flow. Lee *et al.* introduced the so-called plasma rotating arc discharge technique in which the graphite anode is rotated at a high velocity of 10^4 rev min^{-1} [23]. Figure 1.4 shows a schematic diagram of the apparatus. The centrifugal force caused by the rotation generates the turbulence and accelerates the carbon vapor perpendicular to the anode. The yield of the nanotubes can be monitored by changing the rotation speed. Moreover, the rotation distributes the micro discharge uniformly and generates a stable plasma with high temperatures. This enhances anode vaporization, thereby increasing carbon vapor density of nanotubes significantly. Such a technique may offer the possibility of producing nanotubes on the large scale.

Another mass production route of MWNTs can be made possible by generating electric arc discharges in liquid nitrogen [24], as shown in Figure 1.5. Liquid nitrogen prevents the electrode from contamination during arc discharge. The content of

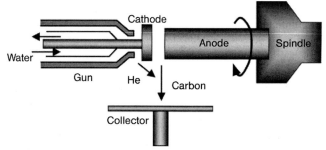

Figure 1.4 Schematic diagram of plasma rotating electrode process system. Reproduced with permission from [23]. Copyright © (2003) Elsevier.

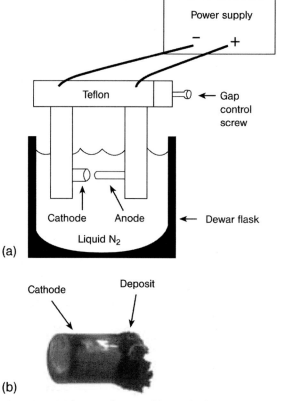

Figure 1.5 (a) Schematic drawing of the arc discharge apparatus and (b) side image of the MWNTs rich material deposited on the cathode. Reproduced with permission from [24]. Copyright © (2003) Springer Verlag.

MWNTs can be as high as 70% of the reaction product. Auger electron spectroscopic analysis reveals that nitrogen is not incorporated in the MWNTs. This technique is considered an economical route for large scale synthesis of high crystalline MWNTs as liquid nitrogen replaces expensive inert gas and cooling system for the cathode.

It is worth noting that SWNTs can only be synthesized through the arc discharge process in the presence of metal catalysts. Typical catalysts include transition metals, for example, Fe, Co and Ni, rare earths such as Y and Gd, and platinum group metals such as Rh, Ru and Pt [25–32]. In this respect, the graphite anode is doped with such metal catalysts. The synthesized nanotubes generally possess an average diameter of 1–2 nm and tangle together to form bundles in the soot, web and string-like structures. The as-grown SWNTs exhibit a high degree of crystallinity as a result of the high temperature of the arc plasma [30]. However, SWNTs contain a lot of metal catalyst and amorphous carbon, and must be purified to remove them. The yield of SWNTs produced from electric arc discharge is relatively low. The diameter and yield of SWNTs can be controlled by using a mixed gaseous atmosphere such as inert–inert or inert–hydrogen mixture [29, 30], and a mixture of metal catalyst particles [28, 32].

1.3.2
Laser Ablation

Laser ablation involves the generation of carbon vapor species from graphite target using high energy laser beams followed by the condensation of such species. The distinct advantages of laser ablation include ease of operation and production of high quality product, because it allows better control over processing parameters. The disadvantages are high cost of the laser Source and low yield of nanotubes produced.

Laser beams are coherent and intense with the capability of attaining very fast rates of vaporization of target materials. In the process, graphite target is placed inside a quartz tube surrounded by a furnace operated at 1200 °C under an inert atmosphere. The target is irradiated with a laser beam, forming hot carbon vapor species (e.g. C_3, C_2 and C). These species are swept by the flowing gas from the high-temperature zone to a conical copper collector located at the exit end of the furnace [33, 34]. Pulsed laser beam with wavelengths in infrared and visible (CO_2, Nd:YAG) or ultravioletUV (excimer) range can be used to vaporize a graphite target. This is commonly referred to as "pulsed laser vaporization" (PLV) technique [35–40]. Moreover, CO_2 and Nd:YAG lasers operated in continuous wave mode have been also reported to produce nanotubes [41–43].

SWNTs can also be produced by laser ablation but require metal catalysts as in the case of electric arc discharge. Figure 1.6 shows a TEM image of deposits formed by the laser (XeCl excimer) ablation of a graphite target containing 1.2% Ni and 1.2% Co [42]. Bimetal catalysts are believed to synthesize SWNTs more effectively than monometal catalysts [33, 36]. From Figure 1.6, SWNTs having a narrow diameter distribution (average diameter of about 1.5 nm) tend to tangle with each other to form bundles or ropes with a thickness of about 20 nm. Their surfaces are coated with amorphous carbon. Metal nanoparticles catalysts with the size of few to 10 nm can be

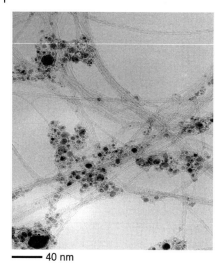

—— 40 nm

Figure 1.6 Transmission electron micrograph of a weblike deposit formed by laser ablation of a graphite target containing 1.2% Ni and 1.2%Co. The metal catalysts appeared as dark spots in the micrograph. Reproduced with permission from [40]. Copyright © (2007) Elsevier.

readily seen in the micrograph. The diameter and yield of SWNTs are strongly dependent on processing parameters such as furnace temperature, chamber pressure, laser properties (energy, wavelength, pulse duration and repetition rate) and composition of target material. The diameter of SWNTs produced by pulsed Nd:YAG laser can be tuned to smaller sizes by reducing the furnace temperature from 1200 °C down to a threshold temperature of 850 °C [35]. Using UV laser irradiation, SWNTs can be synthesized at a lower furnace temperature of 550 °C, which is much lower than the threshold temperature of 850 °C by using Nd:YAG laser [37]. Recently, Elkund et al. employed ultrafast (subpicosecond) laser pulses for large-scale synthesis of SWNTs at a rate of $\sim 1.5\,\text{g h}^{-1}$ [44].

To achieve controlled growth of the SWNTs, a fundamental understanding of their growth mechanism is of particular importance. The basic growth mechanism of SWNTs synthesized by physical vapor depositionPVD techniques is still poorly understood. The vapor–liquid–solid (VLS) mechanism is widely used to describe the catalytic formation of nanotubes via PVD [44, 45] and chemical vapor deposition (CVD) techniques [46, 47]. This model was originally proposed by Wagner and Willis to explain the formation of Si whiskers in 1964 [48]. The whiskers were grown by heating a Si substrate containing Au metal particles in a mixture of $SiCl_4$ and H_2 atmosphere. An Au–Si liquid droplet was formed on the surface of the Si substrate, acting as a preferred sink for arriving Si atoms. With continued incorporation of silicon atoms into the liquid droplets, the liquid droplet became saturated. Once the liquid droplet was saturated, growth occurred at the solid–liquid interface by precipitation of Si from the droplet. Wu and Yang [49] then observed direct formation

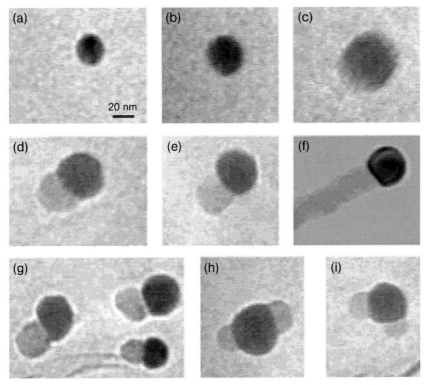

Figure 1.7 *In situ* TEM images recorded during the process of nanowire growth. (a) Au nanoclusters in solid state at 500 °C; (b) alloying initiates at 800 °C, at this stage Au exists in mostly solid state; (c) liquid Au/Ge alloy; (d) the nucleation of Ge nanocrystal on the alloy surface; (e) Ge nanocrystal elongates with further Ge condensation and eventually a wire forms (f); (g) several other examples of Ge nanowire nucleation; (h), (i) TEM images showing two nucleation events on single alloy droplet. Reproduced with permission from [49]. Copyright © (2001) The American Chemical Society.

of Ge nanowire from an Au–Ge liquid droplet using *in situ* TEM (Figure 1.7). Au nanoclusters and carbon-coated Ge microparticles were dispersed on TEM grids, followed by *in situ* heating up to 900 °C. Au nanoclustes remain in the solid state up to 900 °C in the absence of Ge vapor condensation. With increasing amount of Ge vapor condensation, Ge and Au form an eutectic Au–Ge alloy. The formation and growth of Ge nanowire is explained on the basis of VLS mechanism. Figure 1.8 shows schematic diagrams of the nucleation and growth of Ge nanowire from the eutectic Au–Ge liquid alloy.

1.3.3
Chemical Vapor Deposition

The CVD process involves chemical reactions of volatile gaseous reactants on a heated sample surface, resulting in the deposition of stable solid products on

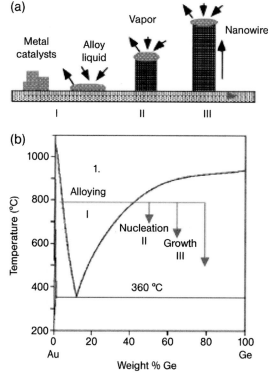

Figure 1.8 (a) Schematic illustration of vapor–liquid–solid nanowire growth mechanism including three stages: (I) alloying, (II) nucleation, and (III) axial growth. The three stages are projected onto the conventional Au–Ge binary phase diagram (b) to show the compositional and phase evolution during the nanowire growth process. Reproduced with permission from [49]. Copyright © (2001) The American Chemical Society.

the substrate. It differs distinctly from PVD techniques that involve no chemical reactions during deposition. CVD has found widespread industrial applications for the deposition of thin films and coatings due to its simplicity, flexibility, and low cost. Moreover, CVD has the ability to produce high purity ceramic, metallic and semiconducting films at high deposition rates. CVD is a versatile and cost-effective technique for CNT synthesis because it enables the use of a feedstock of hydrocarbons in solid, liquid or gas phase and a variety of substrates, and permits the growth of nanotubes in the forms of powder, thin film or thick coating, randomly oriented or aligned tubes. The process involves the decomposition of hydrocarbon gases over supported metal catalysts at temperatures much lower than the arc discharge and laser ablation. The type of CNTs produced in CVD depends on the synthesis temperatures employed. MWNTs are generally synthesized at lower temperatures

(600–900 °C) whereas SWNTs are produced at higher temperatures (900–1200 °C). However, MWNTs prepared by CVD techniques contain more structural defects than those fabricated by the arc discharge. This implies that the structure of CVD-prepared MWNTs is far from the ideal rolled-up hexagonal carbon ring lattice.

Film formation during CVD process includes several sequential steps [50]:

(a) transport of reacting gaseous from the gas inlet to the reaction zone;
(b) chemical reactions in the gas phase to form new reactive species;
(c) transport and adsorption of species on the surface;
(d) surface diffusion of the species to growth sites;
(e) nucleation and growth of the film;
(f) desorption of volatile surface reaction products and transport of the reaction by-products away from the surface.

CVD can be classified into thermal and plasma-enhanced and laser-assisted processes depending upon the heating Sources used to activate the chemical reactions. The heating Sources decompose the molecules of gas reactants (e.g. methane, ethylene or acetylene) into reactive atomic carbon. The carbon then diffuses towards hot substrate that is coated with catalyst particles whose size is of nanometer scale. The catalyst can be prepared by sputtering or evaporating metal such as Fe, Ni or Co onto a substrate. This is followed by either chemical etching or thermal annealing treatment to induce a high density of catalyst particle nucleation on a substrate [46, 47]. Alternatively, metal nanoparticles can be nucleated and distributed more uniformly on a substrate by using a spin-coating method [51, 52]. The size and type of catalysts play a crucial role on the diameter, growth rate, morphology and structure of synthesized nanotubes [53, 54]. In other words, the diameter and length of CNTs can be controlled by monitoring the size of metal nanoparticles and the deposition conditions. Choi et al. demonstrated that the growth rate of CNTs increases with decreasing the grain size of Ni film for plasma-enhanced CVD [54]. As the size of particles decreases, the diffusion time for carbon atoms to arrive at the nucleation sites becomes shorter, thereby increasing the growth rate and density of nanotubes. The yield of nanotubes depends greatly on the properties of substrate materials. Zeolite is widely recognized as a good catalyst support because it favors formation of high yield CNTs with a narrow diameter distribution [55]. Alumina is also an excellent catalyst support because it offers a strong metal-support interaction, thereby preventing agglomeration of metal particles and offering high density of catalytic sites.

1.3.3.1 Thermal CVD

In thermal CVD, thermal energy resulting from resistance heating, r.f. heating or infrared irradiation activates the decomposition of gas reactants. The simplest case is the use of a quartz tube reactor enclosed in a high temperature furnace [48–51, 53]. In the process, the metal catalyst (Fe, Co or Ni) film is initially deposited on a substrate, which is loaded into a ceramic boat in the CVD reactor. The catalytic metal film is treated with ammonia gas at 750–950 °C to induce the formation of metal nanoparticles on the substrate. Hydrocarbon gas is then introduced into the quartz tube reactor [48, 53].

According to the VLS model, an initial step in the CVD process involves catalytic decomposition of hydrocarbon molecules on metal nanoparticles. Carbon atoms then diffuse through the metal particles, forming a solid solution. When the solution becomes supersaturated, carbon precipitates on the surface of particles and grows into CNTs. Growth can occur either below or above the metal catalyst as carbon is precipitated from supersaturated solid solution. Both base and tip growth mechanisms have been suggested. For a strong catalyst–substrate interaction, a CNT grows up with the catalyst particle pinned at its base, favoring a base-growth phenomenon [56, 57]. In the case of a tip growth mechanism, the catalyst–substrate interaction is weak, hence the catalyst particle is lifted up by the growing nanotube such that the particle is eventually encapsulated at the tip of a nanotube [58–60]. Figure 1.9(a) and (b)

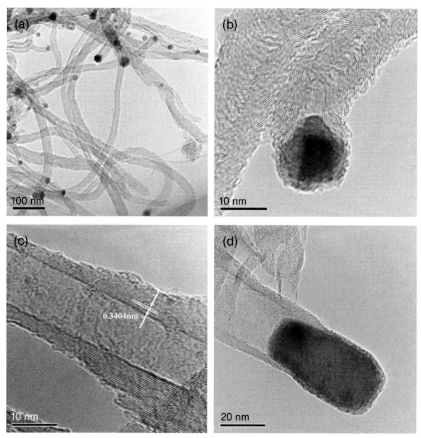

Figure 1.9 TEM images of CNTs synthesized using the catalyst with 10% nickel at 450 °C [(a) and (b)], and at 650 °C [(c) and (d)]. Reproduced with permission from [59]. Copyright © (2008) Elsevier.

show TEM images of MWNTs synthesized from thermal CVD over Ni/Al catalyst at 450 °C [59]. Long nanotubes having a diameter of about 15 nm exhibit typical coiled, spaghetti-like morphology. Some metal catalysts appear to locate at the tip of nanotubes. TEM image of MWNTs produced at 650 °C clearly shows an interlayer spacing between graphitic sheets of 0.34 nm (Figure 1.9(c)). A metal nanoparticle can be readily seen at the tip of the tube formed at 650 °C, thus indicating that a tip growth mechanism prevails in this case (Figure 1.9(d)).

Despite the fact that the thermal CVD method produces tangled and coiled nanotubes, this technique is capable of forming aligned arrays of CNTs by precisely adjusting the reaction parameters [53, 56–58, 61–63]. For practical engineering applications in the areas of field emission display and nanofillers for composites, formation of vertically aligned nanotubes is highly desirable. To achieve this, a gaseous precursor is normally diluted with hydrogen-rich gas, that is, NH_3 or H_2. Lee et al. reported that NH_3 is essential for the formation of aligned nanotubes [61]. Choi et al. [62] demonstrated that the ammonia gas plays the role of etching amorphous carbon during the earlier stage of nucleation of nanotubes. Both synthesis by C_2H_2 after ammonia pretreatment, and synthesis by NH_3/C_2H_2 gas mixture without ammonia treatment favor alignment of nanotubes. Figure 1.10(a) shows the morphology of nanotubes synthesized by flowing C_2H_2 on a Ni film at 800 °C without ammonia pretreatment. Apparently, coiled nanotubes are produced without NH_3 treatment. However, well-aligned nanotubes can be formed on the Ni film with ammonia pretreatment prior to the introduction of C_2H_2 gas (Figure 1.10(b)).

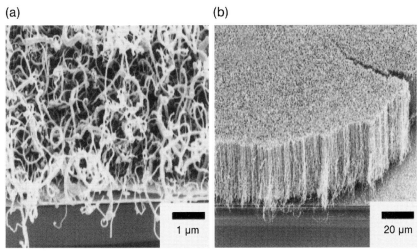

Figure 1.10 Surface morphology of CNTs synthesized by C_2H_2 gas mixture on 20 nm thick Ni films: (a) without and (b) with ammonia treatments. Reproduced with permission from [62]. Copyright © (2001) Elsevier.

1.3.3.2 Plasma-enhanced CVD

Plasma-enhanced CVD (PECVD) offers advantages over thermal CVD in terms of lower deposition temperatures and higher deposition rates. Carbon nanotubes can be produced at the relatively low temperature of 120 °C [64], and even at room temperature [65] using plasma-enhanced processes. This avoids the damage of substrates from exposure to high temperatures and allows the use of low-melting point plastics as substrate materials for nanotubes deposition. The plasma Sources used include direct current (d.c.), radio frequency (r.f.), microwave and electron cyclotron resonance microwave (ECR-MW). These Sources are capable of ionizing reacting gases, thereby generating a plasma of electrons, ions and excited radical species. In general, vertically aligned nanotubes over a large area with superior uniformity in diameter and length can be synthesized by using PECVD techniques [54, 66–70].

Among these Sources, microwave plasma is commonly used. The excitation microwave frequency (2.45 GHz) oscillates electrons that subsequently collide with hydrocarbon gases, forming a variety of C_xH_y radicals and ions. During plasma-enhanced deposition, hydrocarbon feedstock is generally diluted with other gases such as hydrogen, nitrogen and oxygen. Hydrogen or ammonia in hydrocarbon Sources inhibit the formation of amorphous carbons and enhance the surface diffusion of carbon. This facilitates formation of vertically aligned nanotubes. By regulating the types of catalyst, microwave power and the composition ratios of gas precursor, helix-shaped and spring-like MWNTs have been produced [71].

1.3.3.3 Laser assisted CVD

Laser assisted CVD (LCVD) is a versatile process capable of depositing various kinds of materials. In the process, a laser (CO_2) is used to locally heat a small spot on the substrate surface to the temperature required for deposition. Chemical vapor deposition then occurs at the gas–substrate interface. As the spot temperature increases and the reaction proceeds, a fiber nucleates at the laser spot and grows along the direction of laser beam [72]. LCVD generally has deposition rates several orders of magnitude higher than conventional CVD, thus offering the possibility to scale-up production of CNTs. Focused laser radiation permits growth of locally defined nanotube films. Rohmund et al. reported that films of vertically aligned MWNTs of extremely high packing density can be produced using LCVD under conditions of low hydrocarbon concentration [73]. The use of LCVD for producing nanotubes is still in the earlier development stage. Compared with PECVD, only few studies have been conducted on the synthesis of CNTs using LCVD [50, 73, 74].

1.3.3.4 Vapor Phase Growth

This refers to a synthetic process for CNTs in which hydrocarbon gas and metal catalyst particles are directly fed into a reaction chamber without using a substrate. VGCFs can be produced in higher volumes and at lower cost from natural gas using this technique. In 1983, Tibbets developed a process for continuously growing vapor grown carbon fibers up to 12 cm long by pyrolysis of natural gas in a stainless steel tube [75]. However, the diameters of graphite fibers produced are quite large, from 5 to 1000 μm depending on the growth temperature and gas flow rate.

Since then, large efforts have been spent to grow carbon fibers with diameters in the nanometer range (50–200 nm) in the vapor phase by catalytic decomposition of hydrocarbons [17, 76]. Currently, Applied Sciences Inc (ASI, Cedarville, Ohio, USA) produces VGCFs denoted as Pyrograf III of different types, namely PR-1, PR-11, PR-19 and PR-24 [77, 78]. The PR-1, PR-11, and PR-19 nanofibers are 100–200 nm in diameter and 30–100 µm in length. The PR-24 fibers are 60–150 nm in length and 30–100 µm long [77]. Pyrograf III fibers find widespread application as reinforcement materials for polymer composites [6]. In Europe, VGCFs are also commercially available from Electrovac Company [79, 80]. These nanofibers include ENF100 AA (diameter of 80–150 nm), HTF110FF (diameter of 70–150 nm) and HTF150FF (diameter of 100–200 nm); these fibers are longer than 20 µm.

Satishkumar et al. [81] and Jang et al. [82] synthesized CNTs by pyrolysis of iron pentacarbonyl [$Fe(CO)_5$] with methane, acetylene or butane. Ferrocene with a sublimation temperature of ~140 °C is widely recognized to be an excellent precursor for forming Fe catalyst particles. Figure 1.11 is a schematic diagram showing a typical pyrolysis apparatus for the synthesis of CNTs. Ferrocene is placed in the first furnace heated to 350 °C under flowing argon. The ferrocene vapor is carried by the argon gas into the second furnace maintained at 1100 °C. Hydrocarbon gas is finally introduced into the pyrolysis zone of the second furnace. The pyrolysis of ferrocene–hydrocarbon mixture yields iron nanoparticles. After the reaction, carbon deposits are accumulated at the wall of a quartz tube near the inlet end of the second furnace. Rohmund et al. used a single-step, single furnace for the catalytic decomposition of C_2H_2 by ferrocene [83]. In some cases, MWNTs can be produced by pyrolyzing homogeneously dispersed aerosols generated from a solution containing both the hydrocarbon source (xylene or benzene) and ferrocene [84]. Andrews et al. reported that large quantities of vertically aligned MWNTs can be produced through the catalytic decomposition of a ferrocene–xylene mixture at

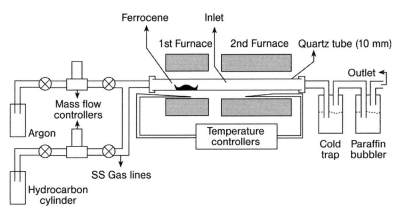

Figure 1.11 Schematic layout of a processing system designed for the synthesis of aligned CNTs by the pyrolysis of ferrocene–hydrocarbon mixtures. Reproduced with permission from [81]. Copyright © (1999) Elsevier.

~675 °C [84]. In engineering practice, a vertical tubular fluidized-bed reactor is generally used to mass-produce CNTs. Fluidization involves the transformation of solid particles into a fluid-like state through suspension in a gas or liquid. A fluidized-bed reactor allows continuous addition and removal of solid particles without stopping the operation. Wang *et al.* developed a nano-agglomerate fluidized-bed reactor in which a high yield of MWNTs of few kilograms per day can be synthesized by continuous decomposition of ethylene or propylene on the alumina supported Fe catalyst at 500–700 °C [85].

1.3.3.5 Carbon Monoxide Disproportionation

In 1999, Smalley's research group at Rice University developed the so-called high pressure CO disproportionation (HiPCo) process that can be scaled up to industry level for the synthesis of SWNTs [86]. SWNTs are synthesized in the gas phase in a flow reactor at high pressures (1–10 atm) and temperatures (800–1200 °C) using carbon monoxide as the carbon feedstock and gaseous iron pentacarbonyl as the catalyst precursor. Solid carbon is produced by CO disproportionation that occurs catalytically on the surface of iron particles via the following reaction:

$$CO + CO \rightarrow C(s) + CO_2 \tag{1.4}$$

Smalley demonstrated that metal clusters initially form by aggregation of iron atoms from the decomposition of iron pentacarbonyl. Clusters grow by collision with additional metal atoms and other clusters, leading to the formation of amorphous carbon free SWNT with a diameter as small as 0.7 nm [86].

In another study, Smalley and coworkers synthesized SWNTs by disproportionation of carbon dioxide on solid Mo metal nanoparticles at 1200 °C [87]. TEM image reveals that the metal nanoparticles are attached to the tips of nanotubes with diameters ranging from 1 to 5 nm. Using a similar approach, researchers at the University of Oklahoma also employed the CO disproportionation process to synthesize SWNTs in a tubular fluidized-bed reactor at lower temperatures of 700–950 °C using silica supported Co–Mo catalysts. They called this synthesis route as the CoMoCat process [88–90]. The bimetallic Co–Mo catalysts suppress formation of the MWNTs but promote SWNTs at lower temperatures. In this method, interactions between Mo oxides and Co stabilize the Co species from segregation through high-temperature sintering. The extent of interaction depends on the Co:Mo ratio in the catalyst [89]. With exposure to CO, the Mo oxide is converted to Mo carbide, and Co is reduced from the oxide state to metallic Co clusters. Carbon accumulates on such clusters through CO disproportionation, leading to the formation of SWNTs. The diameter of SWNTs can be controlled by altering processing conditions such as operating temperature and processing time. However, such synthesized SWNTs contain various impurities such as silica support and the Co and Mo species. Therefore, a further purification process is needed. Despite the improvements that have been achieved in the production of SWNTs, the price and yield of nanotubes are still far from the needs of the industry. Thus, developing a cost-effective technique for producing high yield SWNTs still remains a big challenge for materials scientists.

1.3.4
Patent Processes

Carbon nanotubes with unique, remarkable physical and mechanical characteristics are attractive materials for advanced engineering applications. Scientists from research laboratories worldwide have developed novel technical processes for the synthesis of nanotubes. Table 1.1 lists the patent processes approved by the United States Patent and Trademark Office recently for the synthesis of CNTs.

US Patent 7329398 discloses a process for the production of CNTs or VGCFs by suspending metal catalyst nanoparticles in a gaseous phase [91]. Nanoparticles are prepared in the form of a colloidal solution in the presence or absence of a surfactant. They are then introduced in a gaseous phase into a heated reactor by spraying, injection or atomization together with a carrier and/or a carbon Source. Consequently, most of the problems faced by the conventional gas-phase synthetic processes can be overcome by this novel method.

US Patent 7008605 discloses a non-catalytic process for producing MWNTs by using electric arc discharge technique [92]. The electric current creates an electric arc between the carbon anode and cathode under a protective inert gas atmosphere. The arc vaporizes carbon anode, depositing carbonaceous species on the carbon cathode. The anode and cathode are cooled continuously during discharge. When the electric current is terminated, the carbonaceous residue is removed from the cathode and is purified to yield CNTs.

Table 1.1 Patent processes for production of carbon nanotubes.

Patent Number and Year	Type of Nanotubes	Fabrication Process	Inventor	Assignee
US 7329398 (2008)	MWNTs, VGCF	Vapor phase growth	Kim, Y.N.	KH Chemicals Co. (Korea)
US 7008605 (2006)	MWNTs	Electric arc discharge	Benavides, J.M.	NASA (USA)
US 7094385 (2006)	MWNTs	CVD	Beguin; F., Delpeux; S., Szostak; K.	CNRS (Paris, France)
US 7087207 (2006)	Oriented SWNTs	Laser ablation followed by oxidation and alignment in an electric field	Smalley, R. E., Colbert D.T., Dai, H., Liu, J.; Rinzler, A.G.; Hafner, J.H.; Smith, K., Guo, T., Nikolaev, P., Thess. A.	Rice University (USA)
US 6962892 (2005)	SWNTs	CO disproportionation	Resasco, D.E, Kitiyanan, B., Harwell, J.H., Alvarez, W.E	University of Oklahoma (USA)
US 6994907 (2006)	SWNTs	CO disproportionation	Resasco, D.E, Kitiyanan, B., Harwell, J.H., Alvarez, W.E	University of Oklahoma (USA)

US Patent 7094385 discloses a process for the selective mass production of MWNTs from the decomposition of acetylene diluted with nitrogen on a $Co_xMg_{(1-x)}O$ solid solution at 500–900 °C [93]. Acetylene is a less expensive Source of carbon and easy to use due to its low decomposition temperature, of 500 °C. The process comprises a catalytic step consisting of the formation of nascent hydrogen *in situ* by acetylene decomposition such that CoO is progressively reduced to Co nanometric clusters:

$$C_2H_2 \rightarrow 2C + H_2 \tag{1.5}$$

$$\langle CoO \rangle + H_2 \rightarrow \langle Co \rangle + H_2O \tag{1.6}$$

where $\langle Co \rangle$ represents the Co particles supported on the oxide.

US Patent 7087207 discloses the use of a purified bucky paper of SWNTs as the starting material [94]. Upon oxidation of the bucky paper at 500 °C in an atmosphere of oxygen and CO_2, many tube and rope ends protrude from the surface of the paper. Placing the resulting bucky paper in an electric field can cause the protruding tubes and ropes of SWNTs to align in a direction perpendicular to the paper surface. These tubes tend to coalesce to form a molecular array.

For CO disproportionation, US Patent 6962892 discloses the types of catalytic metal particles used for nanotube synthesis. The bimetallic catalyst contains one metal from Group VIII including Co, Ni, Ru, Rh, Pd, Ir and Pt, and one metal from Group VIB including Cr, W, and Mo. Specific examples include Co–Cr, Co–W, Co–Mo, Ni–Cr, and so on [95]. US Patent 6994907 discloses controlled production of SWNTs based on a reliable quantitative measurement of the yield of nanotubes through the temperature-programmed oxidation technique under CO disproportionation [96]. This permits the selectivity of a particular catalyst and optimizes reaction condition for producing CNTs. The $Co:Mo/SiO_2$ catalysts contain Co:Mo molar ratios of 1:2 and 1:4 exhibit the highest selectivities towards SWNTs.

1.4
Purification of Carbon Nanotubes

1.4.1
Purification Processes

As mentioned above, CNTs prepared from arc discharge, laser ablation and CVD techniques contain various impurities such as amorphous carbon, fullerenes, graphite particles and metal catalysts. The presence of impurities can affect the performance of CNTs and their functional products significantly. Therefore, these impurities must be removed completely from nanotubes using appropriate methods. Typical techniques include gas-phase oxidation [97–99], liquid-phase oxidation [100–106] and physical separation [107–110].

Gas-phase and liquid-phase oxidative treatments are simple practices for removing carbonaceous byproducts and metal impurities from CNTs. In the former case, CNTs are oxidized in air, pure oxygen or chlorine atmosphere at 500 °C [98]. However, oxidative treatment suffers the risk of burning off more than 95% of the nanotube materials [97]. Another drawback of gas-phase oxidation is inhomogeneity of the gas/solid mixture. The liquid-phase oxidative treatment can be carried out simply by dipping nanotubes into strong acids such as concentrated HNO_3, H_2SO_4, mixed 3 : 1 solution of H_2SO_4 and HNO_3, or other strong oxidizing agents such as $KMnO_4$, $HClO_4$ and H_2O_2. In some cases, a two-step (e.g. gas-phase thermal oxidation followed by dipping in acids) or multi-step purification process is adopted, to further improve the purity of CNTs [30, 111–113].

Hiura et al. [100] reported that oxidation of nanotubes in sulfuric or nitric acid is quite slow and weak, but a mixture of the two yields better results. The best oxidant for CNTs is $KMnO_4$ in acidic solution. Hernadi et al. [103] demonstrated that oxidation by $KMnO_4$ in an acidic suspension provides nanotubes free of amorphous carbon. The purification process also opens the nanotube tips on a large scale, leading to the formation of carbonyl and carboxyl functional groups at these sites. The formation of such functional groups is detrimental to the physical properties of CNTs. However, it is beneficial for fully introducing CNTs into polymers. Carbon nanotubes are known to be chemically inert and react very little with polymers. Introducing carboxyl (−COOH) groups into CNTs enhances the dispersion of CNTs in polymers because such functional groups improve the interfacial interaction between them [6]. Chen et al. reported that the efficiency of acid purification of CNTs can be enhanced dramatically by using microwave irradiation [104, 105]. In this technique, MWNTs are initially dispersed in a Teflon container containing 5 M nitric acid solution. The container is then placed inside a commercial microwave oven operated at 210 °C. During the purification process, nitric acid can absorb microwave energy rapidly, thereby dissolving metal particles from nanotubes effectively. More recently, Porro et al. combined acid and basic purification treatment on MWNTs grown by thermal CVD [106]. The purification treatment leads to an opening of the nanotube tips, with a consequent loss of metal particles encapsulated in tips.

Physical separation techniques are based on the initial suspension of SWNTs in a surfactant solution followed by size separation using filtration, centrifugation or chromatography. Bandow et al. reported a multiple-step filtration method to purify laser ablated SWNTs [107]. In this method, the nanotubes were first soaked in a CS_2 solution. The CS_2 insoulubles were trapped in the filter, and the CS_2 solubles (e.g. fullerenes and other carbonaceous byproducts) passed through the filter. The insoluble solids caught on the filter were dispersed ultrasonically in an aqueous solution containing cationic surfactant followed by filtration. This method filters fullerenes and other carbonaceous byproducts, thus the purity of SWNTs is above 90%. The drawback of this method is apparent as it requires several experimental step procedures. Shelimov et al. demonstrated that a single step of ultrasonically assisted filtration can produce SWNT material with purity of >90% and yields of 30–70% [108].

Apart from high purity, length control is another important issue for successful application of CNTs in industry. As mentioned above, CNTs prepared from thermal CVD exhibit coiled morphology with lengths range from several micrometers to

Table 1.2 Comparison of the physical and chemical dispersion methods.

Method	Scale	Mechanism	Main treatment conditions
Shearing	10–100 µm	Breaking up the multi-agglomerates	24 000 r/min, 5 min
Ball milling	10–100 µm	Breaking up the multi-agglomerates	Porcelain ball (20 mm, 11 g), 3.5 h
Ultrasonic	1–300 µm	Dispersing of the single-agglomerates	59 kHz, 80 w, 10 h
Concentrated H_2SO_4/HNO_3	Nano-scale (individual nanotubes)	Dispersing the MWNTs by shortening their length and adding carboxylic groups	3 : 1 concentrated H_2SO_4/HNO_3 ultrasonicated for 0–10 h at 20–40 °C or boiled for 0.5 h at 140 °C

Reproduced with permission from [117]. Copyright © (2003) Elsevier.

millimeter scales. Entangled nanotubes tend to form large agglomerates, thus they must be dispersed or separated into shorter individual tubes prior to the incorporation into composites. The length of CNTs can be reduced through gas-phase thermal oxidation in air at 500 °C and liquid-phase acid purification [114, 115]. Cutting and dispersion of the CNTs can also be induced mechanically via ultrasonication, ball milling, and high speed shearing [116–118]. Wang et al. compared the effects of acid purification, ball milling, shearing, ultrasonication on the dispersion of MWNTs prepared from CVD [117]. They indicated that mechanical treatments can only break up the as-prepared agglomerates of nanotubes into smaller parts of single agglomerates. However, excellent dispersion can be achieved by dipping MWNTs in 3 : 1 H_2SO_4/HNO_3 acid. Their experimental conditions and results are summarized in Table 1.2. It should be noted that the cutting effect of ball milling depends on the type of mill used and milling time. Pierrad et al. demonstrated that ball milling is a good method to obtain short MWNTs (below 1 µm) with open tips [118]. The length of MWNTs decreases markedly with increasing milling time. Table 1.3 lists the current patent processes for the purification and cutting of CNTs.

1.4.2
Materials Characterization

After purification, materials examination techniques must be used to characterize the quality and to monitor material purity of the nanotubes qualitatively or quantitatively. A rapid discrimination of the purity level of CNTs is essential for their effective application as functional materials in electronic devices and structural materials. Analytical characterization techniques used include SEM, TEM, energy dispersed spectroscopy (EDS), Raman, X-ray diffraction (XRD), optical absorption spectroscopy and thermogravimetrical analysis (TGA) [119–121]. SEM and TEM can provide morphological features of purified nanotubes. The residual metal particles present in CNTs can be identified easily by EDS attached to the electron microscopes.

Table 1.3 Patent processes for the purification of carbon nanotubes.

Patent Number and Year	Type of Nanotubes	Purification Process	Inventor	Assignee
US 7135158 (2006)	SWNTs	Multiple-step	Goto, H., Furuta, T., Fujiwara, Y., Ohashi, T.	Honda Giken Kogyo Kabushiki Kaisha (Japan)
US 7029646 (2006)	SWNTs	Fluorination	Margrave, J.L., Gu, Z., Hauge, R.H., Smalley, R.E.	Rice University (USA)
US 7115864 (2006)	SWNTs	Liquid-phase oxidation	Colbert, D.T., Dai, H., Hafner, J.H., Rinzler, A.G., Smalley, R.E., Liu, J., Smith, K.A., Guo, T., Nikolaev, P., Thess, A.	Rice University (USA)
US 6841003 (2005)	SWNTs	Plasma etching	Kang, S.G., Bae, C.	cDream Display Corporation (USA)

Electron microscopy coupled with EDS gives rise to qualitative information for purified nanotubes. In general, TEM observation is preferred because it offers high-resolution images. However, sample preparation for TEM examination is tedious and time consuming. Moreover, electron microscopic observations are mainly focused on localized regions of the samples, thus the analyzed sampling volumes are relatively small.

Raman spectroscopy is a qualitative tool for determining vibrational frequency of molecular allotropes of carbon. All carbonaceous moieties such as fullerenes, CNTs, diamond, and amorphous carbon are Raman active. The position, width and relative intensity of Raman peaks are modified according to the sp^3 and sp^2 configurations of carbon [122–124]. Raman spectra of the SWNTs are well characterized by the low frequency radial breathing mode (RBM) at 150–200 cm^{-1}, with frequency depending on the tube diameter and the tangential mode at 1400–1700 cm^{-1}. From the spectral analysis, a sharp G peak at $\sim$1590 cm^{-1} is related to the (C–C) stretching mode, and a band at $\sim$1350 cm^{-1} is associated with disorder-induced band (D band). Moreover, a second order mode at 2450–2650 cm^{-1} is assigned to the first overtone of D mode and commonly referred to as G' mode. The D-peak is forbidden in perfect graphite and only becomes active in the presence of disorder [122]. In general, the full-width at half-maximum (FWHM) intensity of the D-peak for various carbon impurities is generally much broader than that of the nanotubes. Thus, the purity level of SWNTs can be simply obtained by examining the linewidth of the D-peak [120, 123]. Alternatively, the D/G intensity ratio can be used as an indicator for the purity levels of SWNTs. In the presence of defects, a slight increase in the D-band intensity relative to the intensity of G-band is observed [125].

XRD is a powerful tool for identifying the crystal structure of metal catalysts and carbonaceous moieties. The lattice constant of these species can be determined

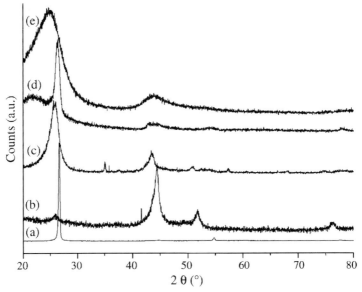

Figure 1.12 X-ray diffraction patterns for (a) graphite; (b) SWNT; (c) MWNT; (d) nanofiber and (e) carbon black. Reproduced with permission from [119]. Copyright © (2006) Elsevier.

accurately from diffraction patterns. It is a fast and non-destructive qualitative tool for routine analysis [120]. Figure 1.12 shows typical XRD patterns for graphite, SWNT, MWNT, carbon nanofiber and carbon black. Pure graphite displays a sharp characteristic peak at 26.6°, corresponding to the (0 0 2) diffracting plane. For SWNT, the (0 0 2) peak becomes broadened and weakened, and the peak position shifts from 26.6° to ~26°. This is attributed to the high curvature and high strain energy resulting from small diameters of SWNTs. Other peaks at 44.5° and 51.7° are also observed, corresponding to the {(1 0 0), (1 0 1)}, and (0 0 4) crystallographic planes. For MWNT, the shape, width and intensity of (0 0 2) peak are modified with the increase in the diameter of the nanotube [119]. The (0 0 2) peak shifts to ~26.2° with respect to that of graphite, and appears slightly asymmetric towards lower angles. The asymmetry is due to the presence of different crystalline species [120, 126].

Ultraviolet-visible-near infrared (UV-vis-NIR) spectroscopy can provide quantitative analysis of the SWNT purity by characterizing their electronic band structures. Because of the one-dimensional structure of SWNTs, the electronic density of state displays several spikes. Accordingly, SWNTs exhibit characteristic electronic transitions associated with Van Hove singularities in the density of states [127]. Semiconducting SWNTs display their first interband transition at S_{11} and second transition at S_{22}, whereas metallic SWNTs show their first transition at M_{11}. Such allowed interband transitions between spikes in the electronic density of states occur in the visible and NIR regions. Figure 1.13 shows a schematic diagram of the absorption

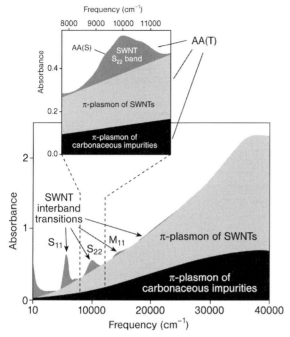

Figure 1.13 Schematic illustration of the electronic spectrum of a typical arc-grown SWNT sample. The inset shows the region of S_{22} interband transition utilized for NIR purity evaluation. In the diagram, AA(S) = area of the S_{22} spectral band after linear baseline correction and AA(T) = total area of the S_{22} band including SWNT and carbonaceous impurity contributions. The NIR relative purity is given by RP = (AA(S)/AA(T))/0.141. Reproduced with permission from [131]. Copyright © (2005) The American Chemical Society.

spectrum of an arc-grown SWNT sample in the spectral range between the far-IR and the UV (10–45 000 cm^{-1}). From the optical absorption spectra of SWNTs in organic solvents such as N,N-dimethylformamide (DMF) or N,N-dimethylacetamide (DMA) over a spectral range, several researchers accurately determined the mass fraction of SWNTs in the carbonaceous portion of a given sample [128–131].

TGA can be used to determine the amount of residual metal catalyst quantitatively in CNTs. Moreover, the thermal behavior of the purified CNTs can be obtained from TGA measurements in air. The thermal stability can be expressed in terms of the peak maximum of mass derivative (dM/dT) from the differential thermogravimetry (DTG) curves. It is well recognized that different structural forms of carbon species exhibit different oxidation behavior or affinity towards oxygen. Thus, carbonaceous impurities of CNTs oxidize in air at different temperatures. In general, CNTs are considerably less stable than graphite in an oxidative environment, particularly SWNTs. However, MWNTs with larger diameters and less strained structures are more thermally stable than SWNTs [119]. Table 1.4 summarizes typical DTG temperatures for various carbonaceous species formed in electric arc-grown SWNTs.

Table 1.4 Reported DTG results for various impurities in arc-grown SWNTs.

Investigator	DTG peak temperature (°C)	Carbonaceous and metal species
Hou et al. [111]	385	Oxidation of metal catalyst
	530	Amorphous carbon, carbon nanoparticle
	830	SWNT
Harutyunyan et al. [132]	~350	Amorphous carbon
	~400	SWNT
	~450	Multi-shell carbon
Shi et al. [133]	364	Amorphous carbon
	434	SWNT
Itkis et al. [131]	349	Amorphous carbon
	418	SWNT

The advantages and drawbacks of the analytical techniques mentioned above for the characterization of purified CNTs are well documented in the literature [119, 120, 131]. Recently, NASA-Johnson Space Center has developed a protocol for the characterization of purified and raw HiPco SWNTs [134]. This protocol summarizes and standardizes analytical procedures in TEM, SEM, Raman, TGA and UV-vis-NIR measurements both qualitatively and quantitatively for characterizing the material quality of SWNTs (Table 1.5).

Table 1.5 A protocol developed by NASA-Johnson Space Center for the characterization of purified and raw HiPco SWNTs.

Parameter	Technique	Analysis
Purity	TGA	Quantitative-residual mass after TGA in air at $5\,°C\,min^{-1}$ to $800\,°C$
	SEM/TEM	Qualitative-amorphous carbon impurities
	EDS	Qualitative-metal content
	Raman	Qualitative-relative amount of carbon impurities and damage/disorder
Thermal stability	TGA	Quantitative-burning temperature in TGA in air at $5\,°C\,min^{-1}$ to $800\,°C$, dM/dT peak maximum
Homogeneity	TGA	Quantitative-standard deviation of burning temperature and residual mass taken on 3–5 samples
	SEM/TEM	Qualitative-Image comparison
Dispersability	Ultra-sonication	Qualitative-time required to fully disperse (to the eye) low conc. SWCNT in DMF using standard settings
	UV-vis-NIR	Quantitative-relative change in absorption spectra of sonicated low concentration SWCNT/DMF solution

Reproduced with permission from [134]. Copyright © (2004) Elsevier.

1.5
Mechanical Properties of Carbon Nanotubes

1.5.1
Theoretical Modeling

The sp^2 carbon–carbon bond of the basal plane of graphite is considered to be one of the strongest in solid materials. Thus, a rolled-up CNT in which all the basal planes run parallel to the tube axis is expected to yield exceptionally high stiff and strong mechanical properties. It is widely recognized that the elastic modulus of a material is closely related to the chemical bonding of constituent atoms. Mutual interatomic interactions can be described by the relevant force potentials associated with chemical bonding. The elastic modulus of covalently bonded materials depends greatly on the interatomic potential energy derived from bond stretching, bending and torsion. In this case, atomic-scale modeling can be used to characterize the bonding, structure and mechanical properties of CNTs.

Theoretical studies of the elastic behavior of CNTs are mainly concentrated on the adoption of different empirical potentials, and continuum mechanics models using elasticity theory [135–139]. Elastic beam models are adopted because CNTs with very large aspect ratios can be considered as continuum beams. These studies show that CNTs are very flexible, capable of undergoing very large scale of deformation during stretching, twisting or bending. This is in sharp contrast to conventional carbon fibers that fracture easily during mechanical deformation.

Yakobson *et al.* used MD simulation to predict the stiffness, and deformation of SWNTs under axial compression [135]. The carbon–carbon interaction was modeled by Tersoff–Brenner potential, which reflects accurately the binding energies and elastic properties of graphite. The predicted Young's modulus value is rather large, at 5.5 TPa. From the plot of strain energy versus longitudinal strain curve, a series of discontinuities is observed at high strains. An abrupt release in strain energy and a singularity in the strain energy vs strain curve are explained in terms of the occurrence of buckling events. Carbon nanotube has the ability to reversibly buckle by changing its shape during deformation. The simulation also reveals that CNTs can sustain a large strain of 40% with no damage to its graphitic arrangement. Thus, CNTs are extremely resilient, and capable of sustaining extremely large strain without showing signs of brittle fracture. Carbon nanotubes only fracture during axial tensile deformation at very high strains of 30% [136]. Iijima *et al.* used computer simulation to model the bending behavior of SWNTs of varying diameters and helicity [137]. They also adopted the Tersoff–Brenner potential for the carbon–carbon interactions. The prediction also shows that CNTs are extremely flexible when subjected to deformation of large strains. The hexagonal network of nanotubes can be retained by bending up to ~110°, despite formation of a kink. The kink allows the nanotube to relax elastically during compression. The simulation is further substantiated by HRTEM observation. From these, it can be concluded that SWNT develops kinks in compression and bending, flattens into deflated ribbon under torsion, and still can reversibly restore its original shape.

Hernandez et al. used a tight-binding model to predict the Young's modulus of SWNTs. The modulus was found to be dependent on the size and chirality of the nanotubes, ranging from 1.22 TPa for the (10,0) and (6,6) tubes to 1.26 TPa for the (20,0) tube [138]. Lu used the empirical force-constant model to determine the elastic modulus of SWNTs and MWNTs [139]. They reported that the elastic modulus of SWNTs and MWNTs are insensitive to the radius, helicity, and the number of walls. Young's modulus of $\sim$1 TPa and shear modulus of $\sim$0.5 TPa were determined from such model. Recently, Natsuki et al. proposed a continuum-shell approach that links molecular and solid mechanics to predict the Young's modulus and shear modulus of SWNTs [140]. SWNT is regarded as discrete molecular structures linked by carbon-to-carbon bonds. Based on this model, the axial modulus was found to decrease rapidly with increasing nanotube diameter. The shear modulus was about half of the Young's modulus. The axial and shear moduli were predicted in the range of 0.61–0.48 TPa, and 0.30–0.27 TPa, respectively.

1.5.2
Direct Measurement

As mentioned above, Yakobson et al. and Iijima et al. reported that CNTs are extremely flexible when subjected to deformation of large strains on the basis of MD simulations [135, 137]. Falvo et al. used atomic force microscopy (AFM) to investigate the bending responses of MWNTs under large strains [141]. They demonstrated that MWNTs can bend repeatedly through large angles using the tip of an atomic force microscope, without suffering catastrophic failure (Figure 1.14(a)–(d)).

Treacy et al. first determined the Young's modulus of arc-grown MWNTs by measuring the amplitude of thermally excited vibrations of nanotubes in a TEM [142]. The nanotubes were assumed equivalent to clamped homogeneous cylindrical cantilevers. TEM images were taken near the tips of free-standing nanotubes. MWNTs were heated from room temperature to 800 °C in steps of 25 °C. Thermal vibrations blurred the image of the tips. The Young's modulus was determined from the slope of the plot of mean-square vibration amplitude (with parameter containing tube length, inner and outer tube diameters) versus temperature. This produced values ranging from 0.4 to 3.70 TPa, with an average value of 1.8 TPa. A large spread in stiffness values arises from the inevitable experimental uncertainties in the estimation of the nanotube length. Another drawback of this method is that the mechanical strength of CNTs cannot be measured. Using the same technique, they obtained a Young's modulus of 1.25−0.35/+0.45 TPa for SWNTs [143].

According to classical continuum mechanics, the elastic modulus of a beam clamped at one or both ends under a given force can yield vertical deflection. Knowing the deflection of a beam, the elastic modulus can be determined from related mathematical equations. Wong et al. employed AFM to determine the elastic modulus and bending strength of arc-grown MWNTs pinned at one end to molybdenum disulfide surface [144]. The nanotubes were then imaged by AFM, allowing the free nanotube to be located. The bending force was measured versus displacement along the unpinned lengths. From the experimental force-displacement curves

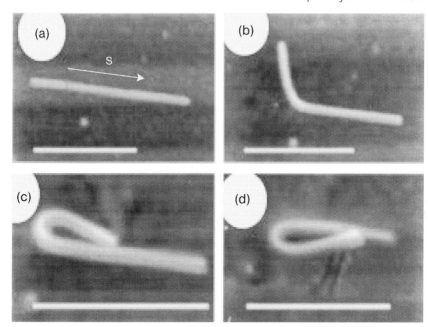

Figure 1.14 Carbon nanotubes in highly strained configuration. The white scale bars at the bottom of each figure represent 500 nm. (a) The original shape of MWNT, 10.5 nm in diameter and 850 nm long. The tube is bent in steps, first upwards (b), until it bends all the way back onto itself (c). The tube is then bent all the way back the other way onto itself (d), to a final curvature similar to that in (c). Reproduced with permission from [141]. Copyright © (1997) Nature Publishing Company.

and mathematical equations based on continuum mechanics, they determined the stiffness to be 1.28 TPa. The bending strength was found to be 14 GPa. Salvetat et al. also used AFM to determine the stiffness of the arc- and CVD- grown MWNTs. The nanotube was clamped at both ends during the measurements [145]. A Si_3N_4 microscope tip was employed to apply the force and for measuring the deflection of the CNT. They obtained an average modulus value of 810 GPa for arc-grown MWNT, and 27 GPa for CVD-MWNT. They attributed the lower stiffness of CVD-MWNT to the formation of defect or structural disorder in the tube. On the basis of TEM observation the graphitic planes are tilted with an angle of 30° with respect to the tube axis. Yu et al. performed in situ tensile measurements in a scanning electron microscope, yielding stiffness values range from 270 to 950 GPa for arc-grown MWNT [146]. They also demonstrated that the MWNT break in the outermost layer, resulting in the morphology commonly described as "sword-in-sheath" feature. The tensile strength of this layer ranges from 11 to 63 GPa.

In general, SWNTs always agglomerate to form ropes with diameters of several tens of nanometers. Using the same equipment, Yu et al. obtained an average Young's modulus of 1 TPa and tensile strength of 30 GPa for SWNT ropes [147]. The tensile strength of SWNT ropes is over 40 times the tensile strength (745 MPa) of typical annealed low alloy steels such as SAE 4340. Low alloys steels are extensively used in

Table 1.6 Experimental Young's modulus and tensile strength of carbon nanotubes synthesized from different techniques.

Investigator	Young's modulus (TPa)	Tensile strength (GPa)	Nanotube type	Synthesis process	Test method
Treacy et al. [142]	1.8	—	MWNT	Electric arc	TEM (thermal vibration)
Krishnam et al. [143]	1.25	—	SWNT	Laser ablation	TEM (thermal vibration)
Wong et al. [144]	1.28	—	MWNT	Electric arc	AFM
Salvetat et al. [145]	0.81	—	MWNT	Electric arc	AFM
Salvetat et al. [145]	0.027	—	MWNT	CVD	AFM
Yu et al. [146]	0.27–0.95	11–63	MWNT	Electric Arc	SEM (tension)
Yu et al. [147]	1	30	SWNT rope	Laser ablation	SEM (tension)
Xie et al. [148]	0.45	3.6	MWNT bundle	CVD	SEM (tension)
Demczyk et al. [149]	0.9	150	MWNT	Electric Arc	TEM (tension)

the manufacture of automotive parts that require high strength and toughness. Xie et al. also determined the stiffness of aligned CVD-MWNT bundle employing a technique similar to Yu et al. [148]. Recently, Demczyk et al. performed *in situ* tension measurement in a TEM for arc-grown MWNT, producing a stiffness of 0.9 TPa and a tensile strength of 0.15 TPa [149]. The results of these investigators are summarized in Table 1.6. From the above discussion, both theoretical and experimental studies confirm that nanotubes have exceptional stiffness and strength. However, a large variation in experimentally measured stiffness and strength is observed, particularly a noticeable reduction in stiffness for CVD-MWNTs when compared with arc-grown MWNTs. This can be expected as these reported values are obtained from a variety of CNTs synthesized from various techniques having different lengths, diameters, disorder structures, level of metal impurities and carbonaceous species.

As mentioned above, the fracture tensile strain of CNTs predicted by the MD simulations could reach up to 30% [136], or even higher depending on simulated temperature and interatomic potential model adopted [150]. Recently, Huang et al. reported that the SWNT fractured at room temperature with a strain $\geq$15% [151]. However, the fracture tensile strain of SWNT could reach up to 280% at temperatures $\geq$2000 °C. (Figure 1.15). In other words, CNT deformed in a superplastic mode at elevated temperatures. In the measurement, a piezo manipulator was used to pull the SWNT to increase the strain under a constant bias of 2.3 V inside a TEM. The temperatures are estimated to be $\geq$2000 °C under a bias of 2.3 V. As a result, kinks and point defects are fully activated in the nanotube, resulting in superplastic deformation.

For VGCFs, the modulus has a strong dependence on the graphite plane misorientation [13]. This is because VGCFs are formed by stacking graphite layers into cone and cup structures. The graphite layers are not parallel to the fiber axis. Accordingly, a wide range of stiffness values have been reported in the literature.

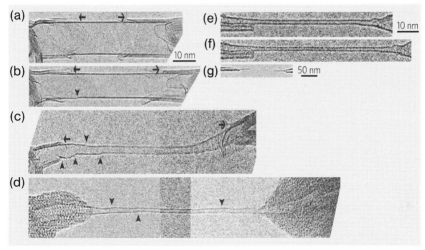

Figure 1.15 *In situ* tensile deformation of individual single-walled CNT in a transmission electron microscope. (a)–(d) Tensile elongation of a SWNT under a constant bias of 2.3 V. Arrowheads mark kinks and arrows indicate features at the ends of the nanotube that are almost unchanged during elongation. Images are scaled to the same magnification. (e), (f) Tensile elongation of a SWNT at room temperature without bias. Images are scaled to the same magnification. (g) Low magnification image showing fracture in midsection of a nanotube at room temperature. Reproduced with permission from [151]. Copyright © (2006) Nature Publishing Company.

Jacobsen *et al.* used the vibration-reed method to measure the stiffness of VGCF and found an average elastic modulus of 680 GPa [152]. Ishioka *et al.* reported that straight VGCFs have an average Young's modulus of 163 GPa and an average tensile strength of 2.05 GPa based on tensile measurements Further, impurities and metal catalysts in VGCFs can result in very low stiffness [153]. Uchida *et al.* calculated the elastic modulus of the two-layered nanofiber to be 100–775 GPa, depending on the degree of misorientation of graphite layers [13].

1.6
Physical Properties of Carbon Nanotubes

1.6.1
Thermal Conductivity

The in-plane thermal conductivity of graphite crystal is very high, that is, ~3080–5150 W m^{-1}K^{-1} at room temperature [154]. However, the c-axis thermal conductivity of graphite is very low because its interlayer is bounded by weak van der Waals forces [155]. The thermal conductivity of graphite generally increases markedly as the temperature is reduced [156], and closely related to their lattice vibrations or phonons. Contribution to a finite in-plane thermal conductivity of graphite at low temperatures is considered to be associated with acoustic phonon scattering. The thermal conductivity of two-dimensional graphite sheet exhibits quadratic temperature dependence at low temperatures. At higher temperatures (say above 140 K), phonon–phonon (umklapp) scattering process dominates.

Berber et al. demonstrated that CNTs have an unusually high thermal conductivity on the basis of MD simulation. The predicted conductivity of SWNT is 6600 W m^{-1}K^{-1} at room temperature [157]. This value is much higher than that of graphite and diamond. The high thermal conductivity of SWNT is believed to result from the large phonon mean free path in one-dimensional nanostructure. The theoretical phonon mean free path length of SWNTs could reach a few micrometers [158]. Considering the umklapp phonon scattering process, Cao et al. developed a model to analyze the thermal transport in a perfect isolated SWNT [159]. They reported that the thermal conductivity of an isolated SWNT increases with the increase of temperature at low temperatures, and shows a peaking behavior at about 85 K before falling off at higher temperatures. The thermal transport in the SWNT is dominated by phonon boundary scattering at low temperatures. As temperature increases, umklapp scattering process becomes stronger and contributes to the heat transfer. Further, SWNTs with small diameters exhibit higher thermal conductivity that is inversely proportional to the tube's diameter at 300 K.

Yie et al. determined the thermal conductivity MWNT bundles using a self-heating 3ω method [160]. The thermal conductivity is approximately linear at temperatures above 120 K, but becomes quadratic at temperatures below that. Subsequently, Kim et al. measured the thermal conductivity of an individual MWNT using a microfabricated suspended device. The observed thermal conductivity is higher than 3000 W m^{-1}K^{-1} at room temperature and the phonon mean free path is ~500 nm [161]. The conductivity value is two orders of magnitude higher than the estimation from previous researchers using macroscopic mat samples (~35 W m^{-1}K^{-1} for SWNTs [162]). Moreover, the measurements shows a nearly T^2-dependence of thermal conductivity for 50 K < T < 150 K. Above 150 K, thermal conductivity deviates from quadratic temperature dependence and displays a peak at 320 K due to the onset of umklapp phonon–phonon scattering. It is noted that the clustering of MWNTs into bundles tends to reduce their thermal conductivity, but has no effect on the temperature dependence of conductivity [148, 158]. When the MWNTs are consolidated into bulk specimens by spark plasma sintering (SPS), the thermal conductivity becomes even lower as a result of the tube–tube interactions [163, 164]. Table 1.7 summarizes the reported thermal conductivity of CNTs synthesized from different processes. Apparently, the thermal conductivity of individual carbon nanotube (3000 W m^{-1}K^{-1}) is significantly higher than silver and copper metals having the highest conductivity of about 400 W m^{-1}K^{-1}.

1.6.2
Electrical Behavior

Because of the anisotropic nature of graphite, the in-plane electrical conductivity of graphite crystal is very high as a result of the overlap of π orbitals of the adjacent

Table 1.7 Theoretical and experimental thermal conductivity at room temperature for carbon nanotubes fabricated from different techniques.

Material	Fabrication process	Measurement technique	Thermal conductivity (W m^{-1} K^{-1})
Individual SWNT [157]	—	MD simulation	6600
Individual MWNT [161]	—	Microfabricated device	3000
MWNT bundle [160]	CVD	Self-heating 3ω method	25
SWNT mat [162]	Arc-discharge	Comparative method	35
Bulk MWNT [163]	SPS (2000 °C, 50 MPa, 3 min)	Laser-flash	4.2
Bulk MWNT [164]	SPS (1700 °C, 50 MPa)	Laser-flash	2.15

carbon atoms. However, the mobility perpendicular to the planes is relatively low [165].

As mentioned before, the electrical properties of CNTs vary from metallic to semiconducting, depending on the chirality and diameter of CNTs. Both theoretical and experimental measurement results demonstrate the superior electrical properties of CNTs. MWNTs can carry extremely high current density from 10^7 to 10^9 A cm^{-2} [166, 167]. The reported room temperature resistivity of individual carbon nanotube is in the order of 10^{-6}–10^{-4} Ω cm [168]. Smalley's group measured the resistivity of metallic SWNT ropes using a four-point technique. The resistivity was found to be ~10^{-4} Ω cm at 300 K [169, 170]. These properties make CNTs the most conductive tubes/wires ever known.

The electrons in one-dimensional CNTs are considered to be ballistically conducted. This implies that the electrons with a large phase coherence length experience no scattering from phonons during ballistic transport in CNTs. Therefore, electrons encounter no resistance and dissipate no heat in CNTs. In this respect, the conductance (the inverse of resistance) of individual SWNTs is predicted to be quantized with a value of $2G_o$, independent of the diameter and the length [171]. The conductance quantum (G_o) can be expressed by the following equation:

$$G_o = 2e^2/h = (12.9 \text{ k}\Omega)^{-1} \tag{1.7}$$

where e is the electronic charge and h is Planck's constant.

Frank et al. measured the conductance of MWNTs using a scanning probe microscope attached with a nanotube fiber. The fiber was carefully dipped into the liquid mercury surface [166]. After contact was established, the current was measured as a function of the position of the nanotubes in the mercury. The results revealed that the nanotubes conduct current ballistically with quantized conductance behavior. The MWNT conductance jumped by increments of one unit of G_o. Recently, Urbina et al. also observed quantum conductance steps of MWNTs in a mixture of polychlorinated biphenyls [172].

Carbon nanotubes with high aspect ratio exhibit excellent field emission characteristics. Electrons can be emitted readily from the nanotube tips by applying a potential between the nanotube surface and an anode. CNTs have been considered as field electron emission Sources for field emission displays (FED), backlights for liquid crystal displays, outdoor displays and traffic signals [173]. However, few FED devices based on CNTs are available commercially because it is difficult to control the properties of CNTs and there is fierce competition with existing organic light-emitting diodes and plasma display panels [174].

1.7
Potential and Current Challenges

A brief overview of the latest state-of-the-art advances in the synthesis, purification, materials characterization, mechanical and physical behavior of CNTs has been presented. It is evident that there is a wealth of synthetic strategies in the development of CNTs but attention is currently focused on both the improvement and simplicity of existing synthetic techniques. Advances in the synthesis of CNTs have continued to improve rapidly in recent years. There are still many challenges to be resolved in their synthesis, purification and cost-effective production processes. Some research efforts are currently underway to overcome these challenges. The price of CNTs, especially SWNTs, is still very expensive when compared with other inorganic fillers commonly used to reinforce metals or ceramics. The price of SWNT is US$ 9990 per 100 g [175]. This has so far restricted their prospects as a substitution for inorganic fillers in composite industries. Furthermore, CNTs synthesized from various techniques contain different levels of metal impurities, amorphous carbon and other carbonaceous byproducts. Purification and subsequent materials examination of nanotubes are costly and time consuming. It is necessary to bring the cost of nanotubes down to an acceptable level through novel innovation of large-scale fabrication and simplified purification processes to manufacture nanotubes of high quality and purity.

Developing high-performance nanocomposites for advanced engineering applications requires the ability to tailor their microstructure, mechanical and physical properties. The successful implementation of such nanocomposites depends on the development of novel processing techniques and fundamental understanding of their structure–property relationship. The development and synthesis of metal- and ceramic nanocomposites is still in the embryonic stage. One of the most important challenges in the fabrication of nanocomposites is how to achieve homogeneous dispersion of nanotubes within a metallic or ceramic matrix. Owing to their tangled feature and large surface area, CNTs tend to disperse as agglomerates or clusters in metals or ceramics during processing. Agglomeration of nanotubes can lead to resulting composites having inferior mechanical strength and toughness as well as poor physical properties. Uniform distribution of nanotubes in metal or ceramic matrix is essential to achieve effective load-bearing capacity of the reinforcement. Therefore, microstructural control is crucial to obtain optimal mechanical and physical properties in metal- and ceramic nanocomposites.

Nomenclature

$\vec{a}_1, \vec{a}_2$ Unit vectors in two-dimensional hexagonal lattice
$\vec{C}_h$ Chiral vector
d Diameter of carbon nanotube
e Electronic charge
G_o Conductance quantum (G_o)
h Planck's constant.
n, m Integers
θ Chiral angle

References

1 Tjong, S.C. (2007) Novel nanoparticle-reinforced metal matrix composites with enhanced mechanical properties. *Advanced Engineering Materials*, **9**, 588–593.

2 Ma, Z.Y., Li, Y.L., Liang, Y., Zheng, F., Bi, J. and Tjong, S.C. (1996) Nanometric Si_3N_4 particulate-reinforced aluminum composite. *Materials Science and Engineering A*, **219**, 229–231.

3 Ma, Z.Y., Tjong, S.C. and Li, Y.L. (1999) The performance of aluminum-matrix composites with nanometric particulate Si-N-C reinforcement. *Composites Science and Technology*, **59**, 263–270.

4 Kang, Y.C. and Chan, S.L. (2004) Tensile properties of nanometric Al_2O_3 particulate-reinforced aluminum matrix composites. *Materials Chemistry and Physics*, **85**, 438–443.

5 Ijima, F S. (1991) Helical microtubules of graphite carbon. *Nature*, **354**, 56–68.

6 Tjong, S.C. (2006) Structural and mechanical properties of polymer nanocomposites. *Materials Science and Engineering R*, **53**, 73–197.

7 Tjong, S.C., Liang, G.D. and Bao, S.P. (2007) Electrical behavior of polypropylene/multi-walled carbon nanotube nanocomposites with low percolation threshold. *Scripta Materialia*, **57**, 461–464.

8 Tjong, S.C., Liang, G.D. and Bao, S.P. (2008) Effects of crystallization on dispersion of carbon nanofibers and electrical properties of polymer nanocomposites. *Polymer Engineering and Science*, **48**, 177–183.

9 Heimann, R.B., Evsyukov, S.E. and Koga, Y. (1997) Carbon allotropes: A suggested classification scheme based on valence orbital hybridization. *Carbon*, **35**, 1654–1658.

10 Kroto, H.W., Heath, J.R., O'Brien, S.C., Curl, R.F. and Smalley, R.H. (1985) C-60-buckminsterfullerene. *Nature*, **318**, 162–163.

11 Rao, C.N., Seshadri, R., Govindaraj, A. and Sen, R. (1995) Fullerenes, nanotubes, onions and related carbon structures. *Materials Science and Engineering R*, **15**, 209–262.

12 Yamabe, T. (1995) Recent development of carbon nanotubes. *Synthetic Metals*, **70**, 1511–1518.

13 Uchida, T., Anderson, D.P., Minus, M.L. and Kumar, S. (2006) Morphology and modulus of vapor grown carbon nanofibers. *Journal of Materials Science*, **41**, 5851–5856.

14 Melechko, A.V., Merkulov, V.I., McKnight, T.E., Guillorn, M.A., Klein, K.L., Lowndes, D.H. and Simpson, M.L. (2005) Vertically aligned carbon nanofibers and related structures:

Controlled synthesis and directed assembly. *Journal of Applied Physics*, **97**, 041301(1)–041301(39).

15 Thostenson, E., Ren, Z. and Chou, T. (2001) Advances in science and technology of carbon nanotubes and their applications: a review. *Composites Science and Technology*, **61**, 1899–1912.

16 Dresselhaus, M.S., Dresselhaus, G. and Saito, R. (1995) Physics of carbon nanotubes. *Carbon*, **33**, 883–891.

17 Tibbets, G.G. (1989) Vapor-grown fibers: status and prospects. *Carbon*, **27**, 745–747.

18 Laplaze, D., Bernier, P., Maser, W.K., Flamant, G., Guillard, T. and Loiseau, A. (1998) Carbon nanotubes: The solar approach. *Carbon*, **36**, 685–688.

19 Ebbesen, T.W. (1994) Carbon nanotubes. *Annual Review of Materials Science*, **24**, 235.

20 Iijima, S., Ajayan, P.M. and Ichihashi, T. (1992) Growth model for carbon nanotubes. *Physical Review Letters*, **69**, 3100–3103.

21 Ijiima, S. (1993) Growth of carbon nanotubes. *Materials Science and Engineering B*, **19**, 172–180.

22 Ebbesen, T.W. and Ajayan, P.M. (1992) Large-scale synthesis of carbon nanotubes. *Nature*, **358**, 220.

23 Lee, S.J., Baik, H.K., Yoo, J.E. and Han, J.H. (2002) Large scale synthesis of carbon nanotubes by plasma rotating arc discharge technique. *Diamond and Related Materials*, **11**, 914–917.

24 Jung, S.H., Kim, M.R., Jeong, S.H., Kim, S.U., Lee, O.J., Lee, K.H., Suh, J.H. and Park, C.K. (2003) High-yield synthesis of multi-walled carbon nanotubes by arc discharge in liquid nitrogen. *Applied Physics A*, **76**, 285–286.

25 Ijima, S. and Ichihashi, T. (1993) Single-shell carbon nanotubes of 1-nm diameter. *Nature*, **363**, 603–605.

26 Bethune, D.S., Kiang, C.H., de Vries, M.S., Gorman, G., Savoy, R., Vasquez, J. and Beyers, R. (1993) Cobalt-catalyzed growth of carbon nanotubes with single-atomic-layer walls. *Nature*, **363**, 603–605.

27 Ajayan, P.M., Lambert, J.M., Bernier, P., Barbedette, L., Colliex, C. and Planeix, J.M. (1993) Growth morphologies during cobalt-catalyzed single-shell carbon nanotube synthesis. *Chemical Physics Letters*, **215**, 509–517.

28 Journet, C. and Bernier, P. (1998) Production of carbon nanotubes. *Applied Physics A*, **67**, 1–9.

29 Saito, Y., Nishikubo, K., Kawabata, K. and Matsumoto, T. (1996) Carbon nanocapsules and single-layered nanotubes produced with platinum-group metals (Ru, Rh, Pd, Os, Ir, Pt) by arc discharge method. *Journal of Applied Physics*, **80**, 3062–3067.

30 Zhao, X., Ohkohchi, M., Inoue, S., Suzuki, T., Kadoya, T. and Ando, Y. (2006) Large-scale purification of single-wall nanotubes prepared electric arc discharge. *Diamond and Related Materials*, **15**, 1098–1102.

31 Farhat, S., de La Chapelle, M.L., Loiseau, A., Scott, C.D., Lefrant, S., Journet, C. and Bernier, P. (2001) Diameter control of single-walled carbon nanotubes using argon-helium mixtures gases. *Journal of Chemical Physics*, **115**, 6752–6759.

32 Lambert, J.M., Ajayan, P.M., Bernier, P., Planeix, J.M., Brontons, V., Coq, B. and Castaing, J. (1994) Improving conditions towards isolating single-shell carbon nanotubes. *Chemical Physics Letters*, **226**, 364–371.

33 Guo, T., Nikolev, P., Thess, A., Colbert, D.T. and Smalley, R.E. (1995) Catalytic growth of single-walled nanotubes by laser vaporization. *Chemical Physics Letters*, **243**, 49–54.

34 Scott, C.D., Arepalli, S., Nikolaev, P. and Smalley, R.E. (2001) Growth mechanisms for single-wall carbon nanotubes in a laser-ablation process. *Applied Physics A*, **72**, 573–580.

35 Bandow, S. and Asaka, S. (1998) Effect of growth temperature on the diameter distribution and chirality of single-wall

carbon nanotubes. *Physical Review Letters*, **80**, 3779–3782.

36 Yudasaka, M., Yamada, R., Sensui, N., Wilkins, T., Ichihashi, T. and Iijima, S. (1999) Mechanism of the effect of NiCo, Ni and Co catalysts on the yield of single-wall carbon nanotubes formed by pulsed Nd:YAG laser ablation. *The Journal of Physical Chemistry B*, **103**, 6224–6229.

37 Dillon, A.C., Parilla, P.A., Alleman, J.L., Perkins, J.D. and Heben, M.J. (2000) Controlling single-wall nanotube diameters with variation in laser pulse power. *Chemical Physics Letters*, **316**, 13–18.

38 Braidy, N., El Khakani, M.A. and Botton, G.A. (2002) Single-wall carbon nanotubes synthesis by means of UV laser vaporization. *Chemical Physics Letters*, **354**, 88–92.

39 Kusaba, M. and Tsunawaki, Y. (2006) Production of single-wall carbon nanotubes by a XeCl excimer laser ablation. *Thin Solid Films*, **506–507**, 255–258.

40 Kusaba, M. and Tsunawaki, Y. (2007) Raman spectroscopy of SWNTs produced by a XeCl excimer laser ablation at high temperatures. *Applied Surface Science*, **253**, 6330–6333.

41 Maser, W.K., Munoz, E., Benito, A.M., Martinez, M.T., de la Fuente, G.F., Maniette, Y., Anglaret, E. and Sauvajol, J.L. (1998) Production of high-density single-walled nanotube material by a simple laser-ablation method. *Chemical Physics Letters*, **292**, 587–593.

42 Bolshakov, A.P., Uglov, S.A., Saveliev, A.V., Konov, V.I., Gorbunov, A.A., Pompe, W. and Graff, A. (2002) A novel CW laser-powder method of carbon single-wall nanotube production. *Diamond and Related Materials*, **11**, 927–930.

43 Dillon, A.C., Parilla, P.A., Jones, K.M., Riker, G. and Heben, M.J. (1998) A comparison of single-wall carbon nanotube production using continuous and pulsed laser vaporization. *Materials Research Society Symposium Proceedings*, **526**, 403–408.

44 Elkund, P.C., Pradhan, B.K., Kim, U.J., Xiong, Q., Fischer, J.E., Friedman, A.D., Holloway, B.C., Jordan, K. and Smith, M.W. (2002) Large-scale production of single-walled carbon nanotubes using ultrafast pulses from a free electron laser. *Nano Letters*, **2**, 561–566.

45 Kukovitsky, E.F., L'vov, S.G. and Sainov, N.A. (2000) VLS-growth of carbon nanotubes from the vapor. *Chemical Physics Letters*, **317**, 65–70.

46 Gruneis, A., Kramberger, C., Grimm, D., Gemming, T., Rummeli, M.H., Barreiro, A., Ayala, P., Pichler, T., Schaman, C., Kuzmany, H., Schumann, J. and Buchner, B. (2006) Eutectic limit for the growth of carbon nanotubes from a thin iron film by chemical vapor deposition of cyclohexane. *Chemical Physics Letters*, **425**, 301–305.

47 Kwok, K. and Chiu, W.K. (2005) Growth of carbon nanotubes by open-air laser-induced chemical vapor deposition. *Carbon*, **43**, 437–446.

48 Wagner, R.S. and Ellis, W.C. (1964) Vapor-liquid-solid mechanism of single crystal growth. *Applied Physics Letters*, **4**, 89–90.

49 Wu, Y. and Yang, P. (2001) Direct observation of vapor-liquid-solid nanowire growth. *Journal of the American Chemical Society*, **123**, 3165–3166.

50 Ohring, F M. (ed) (2002) *Materials Science of Thin Film: Deposition and Structure*, Academic Press, San Diego, USA.

51 Zaretskiy, S.N., Hong, Y.K., Dong, H.H., Yoon, J.H., Cheon, J. and Koo, J.Y. (2003) Growth of carbon nanotubes from Co nanoparticles and C_2H_2 by thermal chemical vapor deposition. *Chemical Physics Letters*, **372**, 300–305.

52 Hong, Y.K., Kim, H., lee, G., Kim, W., Park, J.I., Cheon, J. and Koo, J.K. (2002) Controlled two-dimensional distribution of nanoparticles by spin-coating method. *Applied Physics Letters*, **80**, 844–846.

53 Lee, C.J., Lyu, S.C., Cho, Y.R., Lee, J.H. and Cho, K.I. (2001) Diameter-controlled growth of carbon nanotubes using thermal chemical vapor deposition. *Chemical Physics Letters*, **341**, 245–249.

54 Choi, Y.C., Shin, Y.M., Lee, Y.H., Lee, B.S., Park, G.S., Choi, W.B., Lee, N.S. and Kim, J.M. (2000) Controlling the diameter, growth rate, and density of vertically aligned carbon nanotubes synthesized by microwave plasma enhanced chemical vapor deposition. *Applied Physics Letters*, **76**, 2367–2369.

55 Hernadi, K., Fonseca, A., Nagy, J.B., Bernaerts, D., Fudala, A. and Lucas, A.A. (1996) Catalytic synthesis of carbon nanotubes using zeolite support. *Zeolites*, **17**, 416–423.

56 Choi, G.S., Cho, Y.S., Hong, S.Y., Park, J.B., Son, K.H. and Kim, D.J. (2002) Carbon nanotubes synthesized by Ni-assisted atmospheric pressure thermal chemical vapor deposition. *Journal of Applied Physics*, **91**, 3847–3854.

57 Sohn, J.I., Choi, C.J., Lee, S. and Seong, T.Y. (2002) Effects of Fe film thickness and pretreatments on the growth behaviors of carbon nanotubes on Fe-doped (001) Si substrates. *Japanese Journal of Applied Physics*, **41**, 4731–4736.

58 Ducati, C., Alexandrou, I., Chhowalla, M., Amaratunga, G.A. and Robertson, J. (2002) Temperature selective growth of carbon nanotubes by chemical vapor deposition. *Journal of Applied Physics*, **92**, 3299–3303.

59 Li, H., Zhao, N., He, C., Shi, C., Du, X., Li, J. and Cui, Q. (2008) Fabrication of short and straight carbon nanotubes by chemical vapor deposition. *Materials Science and Engineering A*, **476**, 230–233.

60 Li, H., Shi, C., Du, X., He, C., Li, J. and Zhao, N. (2008) The influences of synthesis temperature and Ni catalyst on the growth of carbon nanotubes by chemical vapor deposition. *Materials Letters*, **62**, 1472–1475.

61 Lee, C.J., Kim, D.W., Lee, T.J., Choi, Y.C., Park, Y.S., Lee, Y.H., Choi, W.B., Lee, N.S., Park, G.S. and Kim, J.M. (1999) Synthesis of aligned carbon nanotubes using thermal chemical vapor deposition. *Chemical Physics Letters*, **312**, 461–468.

62 Choi, K.S., Cho, Y.S., Hong, S.Y., Park, J.B. and Kim, D.J. (2001) Effects of ammonia on the alignment of carbon nanotubes in metal-assisted thermal chemical vapor deposition. *Journal of the European Ceramic Society*, **21**, 2095–2098.

63 Kayastha, V., Yap, Y.K., Dimovski, S. and Gogotsi, Y. (2004) Controlling dissociative adsorption for effective growth of carbon nanotubes. *Applied Physics Letters*, **85**, 3265–3267.

64 Hofmann, S., Ducati, C., Robertson, J. and Kleinsorge, B. (2003) Low-temperature growth of carbon nanotubes by plasma-enhanced chemical vapor deposition. *Applied Physics Letters*, **83**, 135–137.

65 Minea, T.M., point, S., Granier, A. and Touzeau, M. (2004) Room temperature synthesis of carbon nanofibers containing nitrogen by plasma-enhanced chemical vapor deposition. *Applied Physics Letters*, **85**, 1244–1246.

66 Ikuno, T., Honda, S.I., Kamada, K., Qura, K. and Katayama, M. (2005) Effect of oxygen addition to methane on growth of vertically oriented carbon nanotubes by radio-frequency plasma-enhanced chemical vapor deposition. *Journal of Applied Physics*, **97**, 104329 1–104329 4.

67 Eres, G., Puretzky, A.A., Geohegan, D.B. and Cui, H. (2004) *In situ* control of the catalyst efficiency in chemical vapor deposition of vertically aligned carbon nanotubes on predeposited metal catalyst film. *Applied Physics Letters*, **84**, 1759–1761.

68 Bell, M.S., Lacerda, R.G., Teo, K.B., Rupesinghe, N.L., Amaratunga, G.A., Molne, W.I. and Chhowalla, M. (2004) Plasma composition during plasma-enhanced chemical vapor deposition of carbon nanotubes. *Applied Physics Letters*, **85**, 1137–1139.

69 Hsu, C.M., Lin, C.H., Lai, H.J. and Kuo, C.T. (2005) Root growth of multi-wall carbon nanotubes by MPCVD. *Thin Solid Films*, **471**, 140–144.

70 Jang, I., Uh, H.S., Cho, H.J., Lee, W., Hong, J.P. and Lee, N. (2007) Characteristics of carbon nanotubes grown by mesh-inserted plasma-enhanced chemical vapor deposition. *Carbon*, **45**, 3015–3021.

71 Wang, X., Hu, Z., Wu, Q., Chen, X. and Chen, Y. (2001) Synthesis of multi-walled carbon nanotubes by microwave plasma-enhanced chemical vapor deposition. *Thin Solid Films*, **390**, 130–113.

72 Duty, C., Jean, D. and Lackey, W.J. (2001) Laser chemical vapor deposition: materials, modeling, and process control. *International Materials Reviews*, **46**, 271–287.

73 Rohmund, F., Morjan, R., Ledoux, G., Huisken, F. and Alexandrescu, R. (2002) Carbon nanotube films grown by laser-assisted chemical vapor deposition. *Journal of Vacuum Science & Technology B*, **20**, 802–811.

74 Bondi, S.N., Lackey, W.J., Johnson, R.W., Wang, X. and Wang, Z.L. (2006) Laser assisted chemical vapor deposition synthesis of carbon nanotubes and their characterization. *Carbon*, **44**, 1393–1403.

75 Tibbets, G.G. (1983) Carbon fibers produced by pyrolysis of natural gas in stainless steel tubes. *Applied Physics Letters*, **42**, 666–668.

76 Allouche, H., Monthioux, M. and Jacobsen, R.L. (2003) Chemical vapor deposition of pyrolytic carbon on carbon nanotubes. Part 1: Synthesis and morphology. *Carbon*, **41**, 2897–2912.

77 http://www.apsci.com.

78 Tibbetts, G.G., Finegan, J.C., McHugh, J.J., Ting, J.M., Glasgow, D.D. and Lake, M.L. (2000) Applications research on vapor-grown carbon fibers, in *Science and Application of Nanotubes* (eds D. S Tomanek and R.J. S Enbody), Kluwer, New York, USA.

79 http://www.electrovac.com.

80 Hammel, E., Tang, X., Trampert, M., Schmitt, T., Mauthner, K., Eder, A. and Potschke, P. (2004) Carbon nanofibers for composite applications. *Carbon*, **42**, 1153–1158.

81 Satishkumar, B.C., Govindaraj, A. and Rao, C.N. (1999) Bundles of aligned carbon nanotubes obtained by the pyrolysis of ferrocene-hydrocarbon mixtures: role of the metal nanoparticles produced *in situ*. *Chemical Physics Letters*, **307**, 158–162.

82 Jang, J.W., Lee, K.W., Oh, I.H., Lee, E.M., Kim, I.M., Lee, C.E. and Lee, C.J. (2008) Magnetic Fe catalyst particles in vapor phase grown multiwalled carbon nanotubes. *Solid State Communications*, **145**, 561–564.

83 Rohmund, F., Falk, L.K. and Campbell, E.E. (2000) A simple method for the production of large arrays of aligned carbon nanotubes. *Chemical Physics Letters*, **328**, 369–373.

84 Andrews, R., Jacques, D., Rao, A.M., Derbyshire, F., Qian, D., Fan, X., Dickey, E.C. and Chen, J. (1999) Continuous production of aligned carbon nanotubes: a step closer to commercial production. *Chemical Physics Letters*, **303**, 467–474.

85 Wang, Y., Wei, F., Luo, G., Yu, H. and Gu, G. (2002) The large-scale production of carbon nanotubes in a nano-agglomerate fluidized-bed reactor. *Chemical Physics Letters*, **364**, 568–572.

86 Nikolaev, P., Bronikowski, M.J., Bradley, R.K., Rohmund, F., Colbert, D.T., Smith, K.A. and Smalley, R.E. (1999) Gas-phase catalytic growth of single-walled nanotubes from carbon monoxide. *Chemical Physics Letters*, **313**, 91–97.

87 Dai, H., Rinzler, A.G., Nikolaev, P., Thess, A., Colbert, D.T. and Smalley, R.E. (1996) Single-walled nanotubes produced by metal-catalyzed disproportionation of carbon dioxide. *Chemical Physics Letters*, **260**, 471–475.

88 Kitiyanan, B., Alvarez, W.E., Harwell, J.H. and Resasco, D.E. (2000) Controlled production of single-wall carbon nanotubes by catalytic decomposition of CO on bimetallic Co-Mo catalysts. *Chemical Physics Letters*, **317**, 497–503.

89 Herrera, J.E., Balzano, L., Borgna, A., Alvarez, W.E. and Resasco, D.E. (2001) Relationship between the structure/composition of Co-Mo catalysts and their ability to produce single-walled carbon nanotubes by CO disproportionation. *Journal of Catalysis*, **204**, 129–145.

90 Resasco, D.E., Alvarez, W.E., Pompeo, F., Balzano, L., Herrera, J.E., Kitiyanan, B. and Borgna, A. (2002) A scalable process for production of single-walled carbon nanotubes (SWNTs) by catalytic disproportionation of CO on a solid catalyst. *Journal of Nanoparticle Research*, **4**, 131–136.

91 Kim, Y.N. (2008) Preparation of carbon nanotubes. US Patent 7329398.

92 Benavides, J.M. (2006) Method for manufacturing high quality carbon nanotubes. US Patent 7008605.

93 Beguin, F., Delpeux, S. and Szostak, K. (2006) Process for the mass production of multiwalled carbon nanotubes. US Patent 7094385.

94 Smalley, R.E., Colbert, D.T., Dai, H., Liu, J., Rinzler, A.G., Hafner, J.H., Smith, K., Guo, T., Nikolaev, P. and Thess, A. (2006) Method for forming an array of single-wall carbon nanotubes in an electric field and compositions thereof. US Patent 7087207.

95 Resasco, D.E., Kitiyanan, B., Harwell, J.H. and Alvarez, W.E. (2005) Metallic catalytic particle for producing single-walled carbon nanotubes. US Patent 6962892.

96 Resasco, D.E., Kitiyanan, B., Harwell, J.H. and Alvarez, W.E. (2006) Carbon nanotubes product comprising single-walled carbon nanotubes. US Patent 6994907.

97 Ebbesen, T.W., Ajayan, P.M., Hiura, H. and Tanigaki, K. (1994) Purification of nanotubes. *Nature*, **367**, 319–319.

98 Zimmerman, J.L., Bradley, R.K., Huffman, C.B., Hauge, R.H. and Margrave, J.L. (2000) Gas-phase purification of single-walled nanotubes. *Chemistry of Materials*, **12**, 1361–1366.

99 Park, Y.S., Choi, Y.C. and Kim, K.S. (2001) High yield purification of multi-walled carbon nanotubes by selective oxidation during thermal annealing. *Carbon*, **39**, 655–661.

100 Hiura, H., Ebbesen, T.W. and Tanigaki, K. (1995) Opening and purification of carbon nanotubes in high yields. *Advanced Materials*, **7**, 275–276.

101 Rinzler, A., Liu, J., Dai, H., Nikolaev, P., Huffman, C.B., Rodriguez-Macias, F.J., Boul, P.J., Heymann, D., Colbert, D.T., Lee, R.S., Fischer, E., Rao, A.M., Eklund, P.C. and Smalley, R.E. (1998) Large-scale purification of single-walled carbon nanotubes: process, product and characterization. *Applied Physics A*, **67**, 29–37.

102 Chiang, I.W., Brinson, B.E. and Smalley, R.E. (2001) Purification and characterization of single-walled carbon nanotubes. *The Journal of Physical Chemistry. B*, **105**, 1157–1161.

103 Hernadi, K., Siska, A., Thien-Nga, L., Forro, L. and Kiricsi, I. (2001) Reactivity of different kinds of carbon during oxidative purification of catalytically prepared carbon nanotubes. *Solid State Ionics*, **141–142**, 203–209.

104 Chen, C.M., Chen, M., Peng, Y.W., Lin, C.H., Chang, L.W. and Chen, C.F. (2005) Microwave digestion and acidic treatment procedures for the purification of multi-walled carbon nanotubes. *Diamond and Related Materials*, **14**, 798–803.

105 Chen, C.M., Chen, M., Peng, Y.W., Yu, H.W. and Chen, C.F. (2006) High efficiency microwave digestion purification of multiwalled carbon nanotubes synthesized by thermal chemical vapor deposition. *Thin Solid Films*, **498**, 202–205.

106 Porro, S., Musso, S., Vinante, M., Vanzetti, L., Anderle, M., Trotta, F. and Tagliaferro, A. (2007) Purification of carbon nanotubes grown by thermal CVD. *Physica E*, **37**, 58–61.

107 Bandow, S., Rao, A.M., Williams, K.A., Thess, A., Smalley, R.E. and Eklund, P.C. (1999) Purification of single-wall carbon nanotubes by microfiltration. *The Journal of Physical Chemistry. B*, **101**, 8839–8842.

108 Shelimov, K.B., Esenaliev, R.O., Rinzler, A.G., Huffman, C.B. and Smalley, R.E. (1998) Purification of single-walled carbon nanotubes by ultrasonically assisted filtration. *Chemical Physics Letters*, **282**, 429–434.

109 Li, F., Cheng, H.M., Xing, Y.T., Tan, P.H. and Su, G. (2000) Purification of single-walled carbon nanotubes synthesized by the catalytic decomposition of hydrocarbons. *Carbon*, **38**, 2041–2045.

110 Yu, A., Bekyarova, E., Itkis, M.E., Fakhrutdinov, D., Webster, R. and Haddon, R.C. (2006) Application of centrifugation to the large-scale purification of electric arc-produced single-walled carbon nanotubes. *Journal of the American Chemical Society*, **128**, 9902–9908.

111 Hou, P., Liu, C., Tong, Y., Liu, M. and Cheng, H. (2001) Purification of single-walled nanotubes synthesized by the hydrogen arc-discharge method. *Journal of Materials Research*, **16**, 2526–2529.

112 Strong, K.L., Anderson, D.P., Lafdi, K. and Kuhn, J.N. (2003) Purification process for single-walled nanotubes. *Carbon*, **41**, 1477–1488.

113 Montoro, L.A. and Rosolen, J.M. (2006) A multi-step treatment to effective purification of single-walled carbon nanotubes. *Carbon*, **44**, 3293–3301.

114 Ziegler, K.J., Gu, Z., Peng, H., Flor, E.H., Hauge, R.H. and Smalley, R.E. (2005) Controlled oxidative cutting of single-walled carbon nanotubes. *Journal of the American Chemical Society*, **127**, 1541–1547.

115 Li, Q.W., Yan, H., Zhang, J. and Liu, Z.F. (2002) Defect location of individual single-walled carbon, nanotubes with a thermal oxidation strategy. *The Journal of Physical Chemistry. B*, **106**, 11085–11088.

116 Lu, K.L., Lago, L.M., Chen, Y.K., Green, M.L., Harris, P.J. and Tsang, S.C. (1996) Mechanical damage of carbon nanotubes by ultrasound. *Carbon*, **34**, 814–816.

117 Wang, Y., Wu, J. and Wei, F. (2003) A treatment to give separated multi-walled carbon nanotubes with high purity, high crystallization and a large aspect ratio. *Carbon*, **41**, 2939–2948.

118 Pierrard, N., Fonseca, A., Konya, Z., Willems, I., Van Tendeloo, G. and Nagy, J.B. (2001) Production of short carbon nanotubes with open tips by ball milling. *Chemical Physics Letters*, **335**, 1–8.

119 Boccaleri, E., Arrais, A., Frache, A., Giannelli, W., Fino, P. and Camino, G. (2006) Comprehensive spectral and instrumental approaches for the, easy monitoring of features and purity of different carbon nanostructures for nanocomposite applications. *Materials Science and Engineering B*, **131**, 72–82.

120 Belin, T. and Epron, F. (2005) Characterization methods of carbon nanotubes: a review. *Materials Science and Engineering B*, **119**, 105–118.

121 Lafi, L., Cossement, D. and Chahine, R. (2005) Raman spectroscopy and nitrogen vapor adsorption for the study of structural changes during purification of single-wall carbon nanotubes. *Carbon*, **43**, 1347–1357.

122 Ferrari, A.C. and Robertson, J. (2000) Interpretation of Raman spectra of disordered and, amorphous carbon. *Physical Review B-Condensed Matter*, **61**, 14095–14106.

123 Brown, S.D., Jorio, A., Dresselhaus, M.S. and Dresselhaus, G. (2001) Observations of D-band feature in the Raman spectra of, carbon nanotubes. *Physical Review B-Condensed Matter*, **64**, 0734011–0734014.

124 Dresselhaus, M.S., Dresselhaus, G., Jorio, A., Souza Filho, A.G. and Saito, R. (2002) Raman spectroscopy on isolated single wall carbon nanotubes. *Carbon*, **40**, 2043–2061.

125 Dillon, A.C., Parilla, P.A., Alleman, J.L., Gennett, T., Jones, K.M. and Heben, M.J. (2005) Systematic inclusion of defects in pure carbon single-wall nanotubes and their effect on the Raman D-band. *Chemical Physics Letters*, **401**, 522–528.

126 Reznik, D. and Olk, C.H. (1995) X-ray powder diffraction from carbon

nanotubes and, nanoparticles. *Physical Review B-Condensed Matter*, **52**, 116–124.

127 Kataura, H., Kumazawa, Y., Maniwa, Y., Umezu, I., Suzuki, S., Ohtsuka, Y. and Achiba, Y. (1999) Optical properties of single-wall carbon nanotubes. *Synthetic Metals*, **103**, 2555–2558.

128 Itkis, M.E., Perea, D.E., Niyoki, S., Rickard, S.M., Hamon, M.A., Hu, H., Zhao, B. and Haddon, R.C. (2003) Purity evaluation of as-prepared single-walled carbon nanotube soot by use of solution-phase near-IR spectroscopy. *Nano Letters*, **3**, 309–314.

129 Landi, B.J., Ruf, H.J., Evans, C.M., Cress, C.D. and Raffaelle, R.P. (2005) Purity assessment of single-wall carbon nanotubes using, optical absorption spectroscopy. *The Journal of Physical Chemistry. B*, **109**, 9952–9965.

130 Ryabenko, A.G., Dorofeeva, T.V. and Zvereva, G.I. (2004) UV-VIS-NIR spectroscopy study of sensitivity of single-wall carbon nanotubes to chemical processing and Van-der-Waals SWNT/SWNTinteraction. Verification of the SWNT content measurements by, absorption spectroscopy. *Carbon*, **42**, 1523–1535.

131 Itkis, M.E., Perea, D.E., Jung, R., Niyoki, S. and Haddon, R.C. (2005) Comparison of analytical techniques for purity evaluation of single-walled carbon nanotubes. *Journal of the American Chemical Society*, **127**, 3439–3448.

132 Harutyunyan, A.R., Pradhan, B.K., Chang, J., Chen, G. and Eklund, P.C. (2002) Purification of single-wall carbon nanotubes by selective, microwave heating of catalyst particles. *The Journal of Physical Chemistry. B*, **106**, 8671–8675.

133 Shi, Z., Lian, Y., Liao, F.H., Zhou, X., Gu, Z., Zhang, Y., Iijima, S., Li, H., Yue, K.T. and Zhang, S.L. (2000) Large scale synthesis of single-wall carbon nanotubes by arc-discharge method. *Journal of Physics and Chemistry of Solids*, **61**, 1031–1036.

134 Arepalli, S., Nikolaev, P., Gorelik, O., Hadjiev, V.G., Holmes, W., Files, B. and Yowell, L. (2004) Protocol for the characterization of single-wall carbon nanotube material quality. *Carbon*, **42**, 1783–1791.

135 Yakobson, B.I., Brabec, C.J. and Bernholc, J. (1996) Nanomechanics of carbon tubes: Instabilities beyond linear response. *Physical Review Letters*, **76**, 2511–2514.

136 Yakobson, B.I., Campbell, M.P., Brabec, C.J. and Bernholc, J. (1997) High strain rate fracture and C-chain unraveling in carbon nanotubes. *Computational Materials Science*, **8**, 341–348.

137 Iijima, S., Brabec, C., Maiti, A. and Bernholc, J. (1996) Structural flexibility of carbon nanotubes. *Journal of Chemical Physics*, **104**, 2089–2092.

138 Hernandez, E., Goze, C., Bernier, P. and Rubio, A. (1998) Elastic properties of C and $B_xC_yN_z$ composite nanotubes. *Physical Review Letters*, **80**, 4502–4505.

139 Lu, J.P. (1997) Elastic properties of carbon nanotubes and nanoropes. *Physical Review Letters*, **79**, 1297–1300.

140 Natsuki, T., Tantrakarn, K. and Endo, M. (2004) Prediction of elastic properties for single-walled carbon nanotubes. *Carbon*, **45**, 39–45.

141 Falvo, M.R., Clary, G.J., Taylor, R.M. JII, Chi, V., Brooks, F.P. JJr, Wasburn, S. and Superfine, R. (1997) Bending and buckling of carbon nanotubes under large strain. *Nature*, **389**, 582–584.

142 Treacy, M.M., Ebbesen, T.W. and Gibson, J.M. (1996) Exceptionally high Young's modulus observed for individual carbon nanotubes. *Nature*, **381**, 678–680.

143 Krishnan, A., Dujardin, E. and Ebbesen, T.W. (1998) Young's modulus of single-walled nanotubes. *Physical Review B-Condensed Matter*, **58**, 14013–14019.

144 Wong, E.W., Sheehan, P.E. and Lieber, C. (1997) Nanobeam mechanics: elasticity, strength and toughness of nanorods and nanotubes. *Science*, **277**, 1971–1975.

145 Salvetat, J.P., Kulik, A.J., Bonard, J.M., Briggs, A.D., Stockli, T., Metenier, K.,

Bonnamy, S., Beguin, F., Burnham, N.A. and Forro, L. (1999) Elastic modulus of ordered and disordered multi-walled carbon nanotubes. *Advanced Materials*, **11**, 161–165.

146 Yu, M.F., Lourie, O., Dyer, M.J., Kelly, T.F. and Ruoff, S. (2000) Strength and breaking mechanism of multi-walled carbon nanotubes under tensile load. *Science*, **287**, 637–541.

147 Yu, M.F., Files, B.S., Arepalli, S. and Ruoff, S. (2000) Tensile loading of ropes of single wall carbon nanotubes and their mechanical properties. *Physical Review Letters*, **84**, 5552–5555.

148 Xie, S., Li, W., Pan, Z., Chang, B. and Sun, L. (2000) Mechanical and physical properties on carbon nanotube. *Journal of Physics and Chemistry of Solids*, **61**, 1153–1158.

149 Demczyk, B.G., Wang, Y.M., Cumings, J., Hetman, M., Han, W., Zettl, A. and Ritchie, R.O. (2002) Direct mechanical measurement of the tensile strength and, elastic modulus of multiwalled carbon nnaotubes. *Materials Science and Engineering A*, **334**, 173–178.

150 Dereli, G. and Ozdogan, C. (2003) Structural stability and energetics of single-walled carbon nanotubes under uniaxial strain. *Physical Review B-Condensed Matter*, **67**, 0354161–0354166.

151 Huang, J.Y., Chen, S., Wang, Z.Q., kempa, K., Wang, Y.M., Jo, S.H., Chen, G., Dresselhaus, M.S. and Ren, Z.F. (2006) Superplastic carbon nanotubes. *Nature*, **439**, 281–281.

152 Jacobsen, R.L., Tritt, T.M., Guth, J.R., Ehrich, A.C. and Gillespie, D.J. (1995) Mechanical properties of vapor-grown carbon fiber. *Carbon*, **33**, 1217–1221.

153 Ishioka, M., Okada, T. and Matsubara, K. (1992) Mechanical properties of vapor-grown carbon fibers prepared from benzene in Linz-Donawitz converter gas by floating catalyst method. *Journal of Materials Research*, **7**, 3019–3022.

154 Ghosh, S., Calizo, I., Teweldebrhan, D., Pokatilov, E.P., Nika, D.L., Balandin, A.A., Bao, W., Miao, F. and Lau, C.N. (2008) Extremely high thermal conductivity of graphene: Prospects for thermal management applications in nanoelectronic circuits. *Applied Physics Letters*, **92**, 1519111–1519113.

155 Bokros, J.C. (1969) Deposition, structure and properties of pyrolytic carbon, in *Chemistry and Physics of Carbon*, vol. 5 (eds L. S Philip and J.R. S Walker), Marcel Dekker, New York, USA, pp. 97–103.

156 Buerschaper, R.A. (1944) Thermal and electrical conductivity of graphite and carbon at low temperatures. *Journal of Applied Physics*, **15**, 452–454.

157 Berber, S., Kwon, Y.K. and Tomanek, D. (2000) Unusually high thermal conductivity of carbon nanotubes. *Physical Review Letters*, **84**, 4613–4616.

158 Xiao, Y., Yan, X.H., Cao, J.X. and Ding, J.W. (2003) Three-phonon Umklapp process in zigzag single-walled carbon nanotubes. *Journal of Physics: Condensed Matter*, **15**, L341–L347.

159 Cao, J.X., Yan, X.H., Xiao, Y. and Ding, J.W. (2004) Thermal conductivity of zigzag single-walled carbon, nanotubes: Role of the umklapp process. *Physical Review B-Condensed Matter*, **69**, 0734071–0734074.

160 Yi, W., Lu, L., Zhang, D.L., Pan, Z.W. and Xie, S.S. (1999) Linear specific heat of carbon nanotubes. *Physical Review B-Condensed Matter*, **59**, R9015–R9018.

161 Kim, P., Shi, L., Majumdar, A. and McEuen, P.L. (2001) Thermal transport measurements of individual multiwalled nanotubes. *Physical Review Letters*, **87**, 2155021–2155024.

162 Hone, J., Whitney, M., Piskoti, C. and Zettl, A. (1999) Thermal conductivity of single-walled nanotubes. *Physical Review B-Condensed Matter*, **59**, R2514–R2516.

163 Zhang, H.L., Li, J.F., Yao, K.F. and Chen, L.D. (2005) Spark plasma sintering and thermal conductivity of carbon nanotube bulk materials. *Journal of Applied Physics*, **97**, 1143101–1143106.

164 Qin, C., Shi, X., Bai, S.Q., Chen, L.D. and Wang, L.J. (2006) High temperature

electrical and thermal properties of the, bulk carbon nanotube prepared by SPS. *Materials Science and Engineering A*, **420**, 208–211.

165 Krishnan, K.S. and Ganguli, N. (1939) Large anisotropy of the electrical conductivity of graphite. *Nature*, **144**, 667.

166 Frank, S., Poncharal, P., Wang, Z.L. and de Heer, W.A. (1998) Carbon nanotube quantum resistors. *Science*, **280**, 1744–1746.

167 Wei, B.Q., Vajtai, R. and Ajayan, P.M. (2001) Reliability and current carrying capacity of carbon nanotubes. *Applied Physics Letters*, **79**, 1172–1174.

168 Ebbesen, T.W., Lezec, H.J., Hiura, J., Bennett, J.W., Ghaemi, H.F. and Thio, T. (1996) Electrical conductivity of individual carbon nanotubes. *Nature*, **382**, 54–56.

169 Thess, A., Lee, R., Nikolaev, P., Dai, H., Petit, P., Robert, J., Xu, C., Lee, Y.H., Kim, S.G., Rinzler, A.G., Colbert, D.T., cuseria, G.E., Tomanek, D., Fischer, J.E. and Smalley, R.E. (1996) Crystalline ropes of metallic carbon nanotubes. *Science*, **273**, 483–487.

170 Fischer, J.E., Dai, H., Thess, A., Lee, R., Hanjani, N.M., Dehaas, D.L. and Smalley, R.E. (1997) metallic resistivity in crystalline ropes of single-wall carbon, nanotubes. *Physical Review B-Condensed Matter*, **55**, R4921–R4924.

171 Sanvito, S., Kwon, Y.K., Tomanek, D. and Lambert, C.L. (2000) Fractional quantum conductance in carbon nanotubes. *Physical Review Letters*, **84**, 1974–1977.

172 Urbina, A., Echeverria, I., Perez-Garrido, A., Diaz-Sanchez, A. and Abellan, J. (2003) Quantum conductance steps in solutions of multiwalled carbon nanotubes. *Physical Review Letters*, **90**, 1066031–1066034.

173 Choi, W.B., Jin, Y.W., Kim, H.Y., Lee, S.J., Yun, M.J., Kang, J.H., Choi, Y.S., Park, N. S., Lee, N.S. and Kim, J.M. (2001) Electrophoresis deposition of carbon nanotubes for triode-type field emission display. *Applied Physics Letters*, **78**, 1547–1549.

174 Robertson, J. (2004) Realistic applications of CNTs. *Materials Today*, **7**, 46–52.

175 http://www.nanoamor.com/.

2
Carbon Nanotube–Metal Nanocomposites

2.1
Overview

Metal-matrix composites (MMCs) reinforced with ceramic offer the attractive combination of strength, stiffness, wear- and creep-resistant characteristics over monolithic alloys. The composites were initially developed for military and space applications. However, recent demand for materials with specified functional properties has led to their broad applications in ground transportation, automotive, chemical, electronic and recreational industries. Ceramic reinforcements introduced into metal matrices could be continuous fibers, discontinuous short fibers, whiskers or particulates. Continuous ceramic fiber-reinforced MMCs generally possess higher mechanical strength and stiffness, but the high cost of fibers and complicated processing methods make them uneconomical for most industrial applications. In this regard, particulate-reinforced MMCs have received increasingly attention because of their ease of fabrication, lower cost and near-isotropic properties [1]. Table 2.1 lists representative reinforcement materials for MMCs.

In addition to continuous ceramic fiber reinforcement, carbon fibers (CF) can also be used to reinforce metals to form composites with high specific strength and stiffness, low coefficient of thermal expansion, high thermal and electrical conductivity. Carbon fibers are mainly produced from polyacrylonitrile (PAN), and several processing steps are needed to form CFs from this polymeric precursor. These include stabilization, carbonization and graphitization. In the process, PAN precursor solution is initially spun into fibers in which polymer molecular chains align with the fiber direction. These fibers are oxidized at 220 °C under tension, resulting in the fracture of hydrogen bonds and rearrangement of polar nitrile (CN) group into a thermally stable ladder bonding. The stabilized PAN fibers are then pyrolyzed at 1000–1500 °C in an inert atmosphere to form a carbon ring structure. Carbonized fibers with 85–99% carbon content exhibit high tensile strength, low modulus, and very low strain to failure. The graphitization stage is an optional treatment to obtain fibers with high modulus, low tensile strength and extremely low fracture strain. In the process, carbonized fibers are further heated at or above 2000 °C in an inert atmosphere to produce highly oriented graphite layers with carbon content more

Carbon Nanotube Reinforced Composites: Metal and Ceramic Matrices. Sie Chin Tjong
Copyright © 2009 WILEY-VCH Verlag GmbH & Co. KGaA, Weinheim
ISBN: 978-3-527-40892-4

Table 2.1 Typical reinforcement materials for MMCs.

Continuous Fiber	C, B, SiC, Al$_2$O$_3$, Si$_3$N$_4$
Short fiber	SiC, Al$_2$O$_3$
Whisker	TiB, SiC, Si$_3$N$_4$, Al$_2$O$_3$, SiO$_2$
Particulate	SiC, TiC, B$_4$C, Al$_2$O$_3$, TiB$_2$, Si$_3$N$_4$, AlN

than 99%. The physical and mechanical properties of CFs prepared from a PAN precursor are listed in Table 2.2 [2], which shows that commercial grade CFs use a lower cost, modified textile-type PAN. Three different grades of fibers for the aerospace category are derived from different combinations of mechanical stretching, heat treatment and precursor spinning. Carbon fibers can also be prepared from pitch through spinnerets and subsequent heat treatment somewhat similar to those of PAN-based fibers. The pitch Sources include petroleum, coal tar and asphalt. It is noted that the large crystallite size and better orientation of pitch-based fibers give rise to higher modulus, thermal conductivity and lower thermal expansion when compared with PAN-based fibers [3].

Such MMCs are generally fabricated by powder metallurgy (PM) and liquid metal routes. The PM route is mainly used for making discontinuously reinforced MMCs with a near net shape capability The process involves initial mixing of metal powders and ceramic particulates, followed by cold pressing and sintering, or hot pressing/hot

Table 2.2 Properties of PAN-based carbon fibers.

Property	Commercial, standard modulus	Aerospace		
		Standard modulus	Intermediate modulus	High modulus
Tensile modulus/GPa	228	220–241	290–297	345–448
Tensile strength/MPa	380	3450–4830	3450–6200	3450–5520
Elongation at break, %	1.6	1.5–2.2	1.3–2.0	0.7–1.0
Electrical resistivity/$\mu\Omega$ cm	1650	1650	1450	900
Thermal conductivity/W m^{-1}K^{-1}	20	20	20	50–80
Coefficient of thermal expansion, axial direction, 10^{-6} K	−0.4	−0.4	−0.55	−0.75
Density, g cm^{-3}	1.8	1.8	1.8	1.9
Carbon content, %	95	95	95	+99
Filament diameter, μm	6–8	6–8	5–6	5–8
Manufacturers	Zoltek, Fortafil, SGL	BPAmoco, Hexcel, Mitsubishi Rayon, Toho, Toray, Tenax, Soficar, Formosa		

Reprinted from [2]. Copyright © (2001) ASM International.

isostatic pressing. In certain cases, secondary mechanical treatment such as extrusion or forging is required to shape the compacts into full-density products. The main drawback of the PM route is the high cost of raw powders. The liquid metal route offers high production yield and low cost, but suffers from filler segregation and pore formation. Two liquid processing techniques are currently used to prepare aluminum composites reinforced with CFs, that is, liquid metal infiltration and squeeze casting [4]. Melt infiltration involves infiltration of molten aluminum into a fiber preform in a protective nitrogen atmosphere. However, carbon tends to react with aluminum and forms a brittle Al_4C_3 phase due to the high temperature environment, long processing and solidification times. This leads to poor mechanical properties of the resulting composites. A deleterious chemical reaction between carbon and aluminum, poor wetting of carbon by molten aluminum and formation of an intermetallic phase are the main issues encountered in the processing of the Al/CF composites by liquid metal processing.

2.2
Importance of Metal-Matrix Nanocomposites

Carbon fibers display high tensile strength and stiffness, but exhibit extremely low fracture strains. The incorporation of a large volume fraction of rigid CFs into metals deteriorates their tensile ductility and toughness significantly. In this regard, carbon nanotubes (CNTs) with even higher tensile strength, stiffness and flexibility are far superior to CFs as reinforcements for metals. The addition of low loading level of CNTs to metals enhances the strength and stiffness of nanocomposites. Incorporation of CNTs does not seem to impair the fracture toughness of metals. In certain cases, a dramatic improvement in tensile ductility in CNT-reinforced nanocomposites over unreinforced metals has been reported [5]. Furthermore, CNTs with excellent thermal conductivity are effective heat dissipating fillers for metals for producing thermal management components in electronic devices. At present, ceramic particle-reinforced Al-based MMCs are widely used in electronic packaging and thermal management applications. However, extremely large volume contents (>50%) of ceramic particulates are needed to achieve the required thermal dissipating effect [6, 7].

Recently, the escalation of fossil fuel prices and the need to minimize carbon dioxide emission have driven the search for light-weight structural materials in aerospace and automotive industries for the reduction of fuel consumption. Conventional particle-reinforced MMCs for structural applications contain ceramic fillers up to 25 vol% [8, 9]. Such a large amount of microfiller increases the weight of composites and degrades their processability and mechanical performance significantly. The size of ceramic particles generally range from a few to several hundred micrometers. Fracture of microparticles occurs readily during mechanical deformation, with cracks perpendicular to the applied load in tension and parallel to it in compression [10–12]. Therefore, microparticle-reinforced MMCs always possess low tensile strength and ductility. To minimize or avoid particle cracking,

Table 2.3 Tensile properties of PM Al and its composites.

Materials	Yield strength (MPa)	Tensile strength (MPa)	Elongation (%)
Pure Al	68	103	—
Al/1vol.% Si_3N_4 (15 nm)	144	180	17.4
Al/15 vol% SiC (3.5 μm)	94	176	14.5

Reproduced with permission from [13]. Copyright © (1996) Elsevier.

it is necessary to reduce the size of reinforcing ceramic particulates from micrometer to nanometer level. Much effort has been directed towards the development of high performance composites with high strength and modulus by adding ceramic nanoparticles. Tjong and coworkers [13] demonstrated that the tensile strength of pure aluminum increases markedly by adding only 1 vol% Si_3N_4 (15 nm) nanoparticles. The tensile strength of such a nanocomposite is slightly higher than that of aluminum composite reinforced with 15 vol% SiC (3.5 μm) particles. Furthermore, the yield stress of Al/1 vol% Si_3N_4 nanocomposite is significantly higher than that of the Al/15 vol% SiC_p microcomposite. The enhancement in yield strength of the Al-based nanocomposite is attributed to Orowan strengthening mechanism. Moreover, the tensile elongation of nanocomposite is higher than that of microcomposite (Table 2.3). Apparently, reducing the size of ceramic microparticles to nanometer level is very effective in avoiding particle cracking, thereby improving the tensile ductility of nanocomposite. A similar beneficial effect of ceramic nanoparticle additions in improving the tensile performances of magnesium has also been reported in the literature [14, 15]. It is considered that the incorporation of CNTs with larger aspect ratio, higher tensile strength and stiffness as well as better flexibility, into metals can yield even much better mechanical performances than ceramic nanoparticles.

2.3
Preparation of Metal-CNT Nanocomposites

Limited research has been conducted in the preparation, structural, physical and mechanical properties of metal-CNT nanocomposites due to the lack of appropriate processing technique, difficulty of dispersing CNTs in metal matrices and lack of proper understanding of the interfacial issue between CNTs and metals. As mentioned before, CVD-grown MWNTs with large aspect ratios exhibit spaghetti-like coiled features, and SWNTs often agglomerate into ropes of several tens of nanometers. Accordingly, CNTs tend to form clusters in metals during processing, leading to poor mechanical properties. For instance, Kuzumaki *et al.* studied the microstructure and tensile properties of Al/5 vol% CNT and Al/10 vol% CNT nanocomposites prepared by conventional powder mixing, hot pressing followed by hot extrusion [16]. Very little improvement in tensile strength was found in the two composites due to inhomogeneous dispersion of CNTs in the metal matrix.

This implies the absence of effective load transfer effect from metal matrix to CNTs during tensile deformation.

In addition to homogeneous dispersion of nanotubes, the interface between the reinforcement and the matrix plays a crucial role in load transfer and crack deflection mechanisms. A controlled interface with strong CNT-matrix interactions generally promotes effective load transfer effect. More recently, Ci *et al.* investigated the interfacial reaction between MWNTs and aluminum by sputtering pure Al on the surface of aligned MWNT arrays The thin film samples were then annealed at 400–950 °C [17]. Transmission electron microscopyTEM examination revealed that aluminum carbide (Al_4C_3) was formed at the MWNT-Al interfaces and the tips of the MWNTs after annealing. These carbide particles improve the interfacial bonding between CNTs and Al layers. As recognized, processing conditions can affect the structure and property of the nanotube-matrix interface. Generally CNTs have poor compatibility and wetting with metallic materials. Special surface treatments for CNTs [18, 19] and/or modification of conventional processing techniques are adopted by some researchers to achieve stronger interfacial bonding between CNTs and metals.

Another important factor to be considered in fabricating CNT–metal nanocomposites is selection of the correct CNT materials. In general, the use of SWNTs as reinforcements is preferred and beneficial for mechanical properties due to their higher mechanical strength. From the economical viewpoint, MWNTs and VGCFs offer distinct advantages over SWNTs. However, the reinforcement efficiency of MWNTs is far from the expected value. The use of MWNTs as intrinsic reinforcements for composite materials may not allow the maximum strength to be achieved due to non-uniform axial deformation inside the tubes [20]. MWNTs are made up of concentric graphene sheets, linked only by weak van der Waals forces. All the inner sheets can rotate and slide freely at low applied forces, thus can detach from outer walls in a "sword and sheath" mechanism

At present, CNT–metal nanocomposites can be processed by means of thermal spray forming, liquid metal processing, powder metallurgy, severe plastic deformation, friction stir processing, electrochemical deposition and molecular-level mixing techniques. The beneficial effects and drawbacks of these techniques in forming MMCs are discussed in the following sections.

2.4
Aluminum-Based Nanocomposites

2.4.1
Spray Forming

Aluminum and its alloys have been selected as the matrix materials for nanocomposites due to their low melting point, low density, adequate thermal conductivity, heat treatment capability, processing versatility and low cost. Thermal spraying is an effective processing technique for the deposition of a wide variety of materials coatings ranging from metals to ceramics. Thermal spray coatings find widespread applications in aerospace turbines, automotive engines, orthopedic prostheses and electronic devices. In the process, the coating material is heated in a gaseous medium at elevated temperatures and projected at high velocity onto a component surface from the spray gun. Upon impact, molten splats become flattened, transfer their heat to the cold substrate and solidify in a short period of time. Several processes such as plasma spraying, high velocity oxyfuel spraying (HVOF), electric arc spraying and detonation flame spraying can be used to form coatings depending on material and desired coating performance. Synthesis of nanocomposite coatings can also be realized via plasma spraying and HVOF [21]. In plasma spray forming (PSF), an arc is formed between a central tungsten cathode and copper nozzle (anode) in which an inert gas flow is passed through the arc, forming a high temperature plasma jet (~15 000 °C) from the nozzle. The coating material is fed into the plasma jet by a carrier gas, where it is melted and propelled from the gun as a stream of molten droplets at velocities of ~400–800 ms^{-1} (Figure 2.1(a)). To minimize oxidation of the

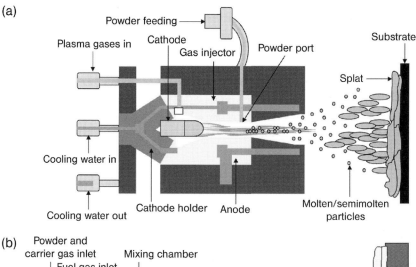

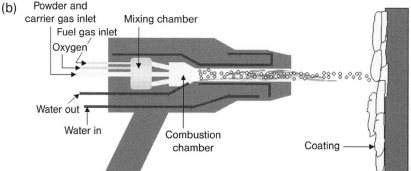

Figure 2.1 Schematic diagrams of (a) plasma spraying and (b) high velocity oxyfuel spraying. Reproduced with permission from [21]. Copyright © (2006) Elsevier.

coating material, the plasma gun and component are enclosed in a low pressure chamber containing an inert atmosphere. Plasma spraying offers improved coating adhesion and higher density of coating due to its high spraying velocity. The HVOF method employs a combustion flame of compressed fuel gas and oxygen to heat the coating material rapidly in an internal combustion at 3000–6000 °C and accelerate the coating material to high velocities ($\sim$700–1800 ms^{-1}) (Figure 2.1(b)). Molten splats encounter extremely high cooling rates during solidification, that is, 10^6–10^8 K s^{-1} for PSF and 10^3–10^5 for HVOF. Thermal spray coatings generally contain pores and consolidation treatment such as sintering, spark plasma sintering, hot pressing or laser irradiation is needed to remove porosities [22].

Recently, Agarwal and coworkers used both plasma spraying and HVOF techniques to fabricate Al-23 wt% Si/10 wt% MWNT nanocomposite coatings (with 0.64 and 1.25 mm thicknesses) on the aluminum substrate [23–26]. Gas atomized Al-23 wt% alloy powder (15–45 µm) was blended and mixed with 10 wt% MWNTs in a ball mill for 48 h to achieve homogeneous mixing prior to spraying. The microstructures of thermally sprayed coatings are heterogeneous, consisting of splats and pores at micro-level, and ultrafine primary Si particles and CNTs at submicron/nanolevel. Moreover, CNTs are located between the splat interfaces and also embedded in the matrix [25]. Figure 2.2(a) and (b) show cross-sectional SEM micrographs of the PSF and HVOF sprayed 10 wt% MWNT/Al nanocomposite coatings. It can be seen that the plasma sprayed coating is more porous, but HVOF sprayed coating is denser due to its higher spraying velocity. TEM examination of the two coatings reveals the presence of nanosized α-Al grains with uniformly distributed ultrafine primary silicon particles [25]. The size of α-Al nanograins for the PSF and HVOF coatings is 87 nm and 64 nm, respectively. The formation of α-Al nanograins is attributed

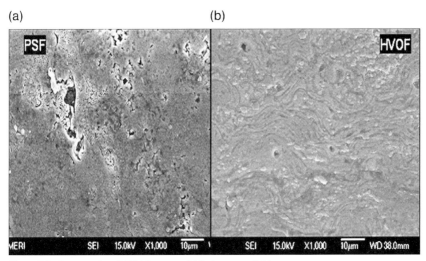

Figure 2.2 Cross-sectional SEM images of (a) plasma and (b) HVOF sprayed MWNT/Al nanocomposite coatings. Reproduced with permission from [26]. Copyright © (2008) Elsevier.

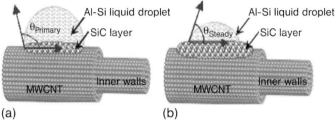

Figure 2.3 Reactive wetting kinetics at MWNT and hypereutectic Al-Si alloy, showing gradual improvement in wettability with formation of SiC layer. Reproduced with permission from [24]. Copyright © (2007) Elsevier.

to the fragmentation of dendritic structure under the heavy impact of spayed molten droplets and restriction of grain growth as a result of very high cooling rates during solidification. The primary silicon nanoparticles are produced by diffusing out of silicon from the hypereutectic Al-Si alloy during solidification.

The wetting behavior between molten Al and the CNTs could lead to the generation of new interface layer. During the initial stage of interfacial reaction, C from MWNTs reacts with Si atoms of Al-Si droplet to form a thin β-SiC layer on the MWNT surface. The contact angle ($\theta_{primary}$) is large at this stage as the Al-Si liquid droplet is in contact with MWNT surface. The SiC layer then spreads and grows laterally until the steady-state reaction is achieved. The contact angle (θ_{steady}) then decreases as the Al-Si alloy flows over the SiC layer (Figure 2.3). TEM examination reveals the formation of β-SiC layer at the nanotube-matrix interface. Plasma sprayed coating exhibits a thicker SiC layer (~5 nm) than the HVOF nanocomposite (~2 nm) (Figure 2.4(a) and (b)). The formation of β-SiC interfacial layer is responsible for the improved wettability of the molten Al-Si alloy matrix with MWNTs [25].

The high strength and stiffness, and excellent flexibility of CNTs make them excellent fillers to improve the wear resistance of bulk aluminum or coatings [19, 27].

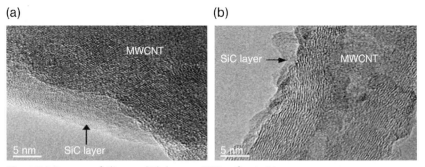

Figure 2.4 Presence of silicon carbide layer on CNT surface in (a) plasma and (b) HVOF formed hypereutectic Al-Si composite reinforced with MWCT. Reproduced with permission from [24]. Copyright © (2007) Elsevier.

Table 2.4 Changes in the size and volume fraction of the porosity and primary silicon in thermally sprayed Al/MWNT nanocomposites after sintering at 400 °C for different time periods.

	As sprayed		24 h sintering at 400 °C		72 h sintering 400 °C	
	PSF	HVOF	PSF	HVOF	PSF	HVOF
Pore						
Size(μm)	1.7–7.8	1.1–2.4	0.8–5.3	1.2–2.1	0.9–4.3	1.0–2.8
%V_{Pore}	6.7 ± 0.4	3.2 ± 0.3	4.3 ± 0.5	2.5 ± 0.2	3.2 ± 0.3	2.1 ± 0.03
Primary Si						
Size(μm)	0.8–2.0	1.2–2.4	0.8–3.3	1.5–4.2	1.1–2.7	0.9–3.5
%V_{Si}	14.1 ± 1.9	14.9 ± 1.0	17.3 ± 1.3	18.3 ± 1.1	17.7 ± 1.5	18.1 ± 1.5

Reproduced with permission from [26]. Copyright © (2008) Elsevier.

However, the presence of pores in plasma- and HVOF sprayed coatings has deleterious effects on the hardness and stiffness. This in turn leads to lower tribological performance. In this regard, post-spraying sintering is needed and found to be very effective in removing pores in plasma and HVOF sprayed coatings (Table 2.4). Moreover, sintering treatments also increase the volume fraction of primary Si particles for both nanocomposite coatings.

In a recent study, Agarwal and coworkers also attempted to deposit Al-12Si/MWNT coatings with thickness up to 500 μm using cold spraying [28]. Cold spraying is an emerging coating process involving the injection of powder particles into a supersonic speed gas jet onto a substrate through a de-Laval type of nozzle. The coating is formed through plastic deformation of sprayed particles in the solid state during impact at temperatures well below the melting point of spray materials. Thus the deleterious effects inherent in conventional thermal spraying processes such as phase transformation, grain growth and oxidation can be minimized or avoided. From microstructural observation, CNTs are also found at the splat interfaces of the coatings. The nanotubes are shortened in length due to the high velocity impact of the spray particles.

2.4.2
Powder Metallurgy Processing

Powder metallurgy processing is a versatile method commonly used to make Al–CNT nanocomposites due to its simplicity. Aluminum alloy powders of micro- or nanometer sizes are blended ultrasonically with CNTs in an organic solvent (e.g., alcohol), followed by solvent evaporation, sintering or hot consolidation of powdered mixture. However, materials scientists have encountered a number of difficulties during PM processing of metal-matrix nanocomposites. For instance, sintering and/or consolidation of nanosized metal powders at high temperatures always lead to fast grain growth. The size of matrix grains can grow up to micrometer scale, depending upon processing temperature and time. In general, it is advantageous to retain the size of matrix grains of composites in nanometer region because significant improvement in hardness, yield strength and wear resistance of composites can be obtained in the presence of nanosized grains. For instance, Zhong et al. attempted to prepare nano-Al/SWNT composites having a nanograined matrix [29]. The composites were prepared by mixing Al nanopowders (53 nm) and purified SWNTs in alcohol ultrasonically. The blended material was dried, cold compacted into disks at room temperature, followed by hot consolidation at 260–480 °C under a pressure of 1 GPa. Excessive grain growth occurred during hot compression of Al nanopowders. The grain size of the matrix of composite grew up to ∼800 nm during hot compaction at 480 °C. The nanotubes have little effects in restricting the grain growth of matrix grains. Furthermore, SWNTs dispersed as bundles rather than individual nanotubes in the Al matrix.

Another complication arising from PM processing of metal-matrix nanocomposites is agglomeration of CNTs. Homogeneous dispersion of nanotubes in Al matrix cannot be achieved by conventional mixing, pressing, sintering of Al powders and CNTs [16]. Effort has been made toward uniform dispersion of CNTs in the metal matrix by modifying conventional PM process. For example, Esawi and El Borady used a powder can rolling technique to fabricate the Al/MWNT nanocomposites [30]. In the process, Al powders and MWCNTs were mixed in a planetary mill for 5 h without using milling media. The mixture was encapsulated in copper cans under an argon atmosphere, and subjected to hot rolling (Figure 2.5). The rolled cans were then sintered in a vacuum furnace at 300 °C for 3 h. The cans were removed, and the strips were sintered in an air furnace at 550 °C for 45 min. This technique can yield better dispersion of CNTs in aluminum matrix by adding very low MWCNT content, that is, 0.5 wt% (Figure 2.6(a) and (b)). Spherical dimples associated with ductile fracture can be readily seen in these micrographs. At 1 wt% MWCNT, nanotubes tend to agglomerate into clusters, leading to inferior mechanical properties (Figure 2.7).

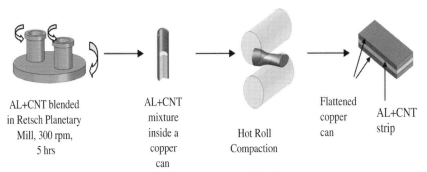

Figure 2.5 Powder blending and can rolling process. Reproduced with permission from [30]. Copyright © (2008) Elsevier.

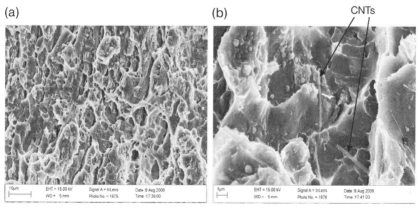

Figure 2.6 (a) Low and (b) high magnification SEM fractographs of Al./0.5wt% MWNT strip. Several individual nanotubes can be seen in (b). Reproduced with permission from [30]. Copyright © (2008) Elsevier.

For metal-matrix microcomposites, the relative ratio of the matrix particle size to reinforcement particle size (RPS) is used as an indicator for homogeneous dispersion of microparticles in the metal matrix. In other words, the degree of homogeneity is expressed in terms the RPS ratio. Higher RPS ratio favors particle clustering whereas low RPS ratio facilitates homogeneous distribution of ceramic microparticles in the metal matrix [31–33]. It is considered that a relatively large RPS ratio between metal powders and CNT particles promotes the formation of nanotube clusters in the CNT-reinforced metals. Accordingly, fabrication methods based on PM have overwhelmingly concentrated on improving nanotube dispersion in order to achieve desired mechanical characteristics in the resulting nanocomposites [34–41].

More recently, dispersion of CNTs homogeneously in metals has been achieved by employing mechanical alloying (MA). This technique is used widely to obtain

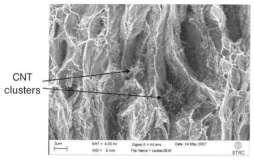

Figure 2.7 SEM fractograph of Al/1 wt% MWNT strip showing nanotube clusters. Reproduced with permission from [30]. Copyright © (2008) Elsevier.

homogeneous dispersion of ceramic microparticles in metals [42–45]. Since MA is a non-equilibrium process, it produces a variety of supersaturated solid solutions, metastable crystallites and amorphous metal alloys. The process involves loading the powders into a high-energy ball mill containing a grinding medium such as stainless steel or alumina balls. The total milling energy can be manipulated by varying the charge ratio (ratio of the weight of balls to the powder), ball mill design, milling atmosphere, time, speed and temperature. Various types of ball mills consisting of vials/tanks and grinding balls can be used for this purpose. These include the Spex shaker mill, platenary ball mill, and attritors [44]. For the shaker mill, the motor vigorously shakes the vials, resulting in high energy impacts between the balls and powder materials. A platenary ball mill employs strong centrifuge forces to develop high-energy milling action inside the vial. In general, large quantity of powders can be processed using an attritor mill consisting of a tank container, rotating impeller and grinding ball.

During mechanical alloying, the powders experience repeated fracturing, welding and plastic deformation as a result of high-energy collision of powders with the ball media during milling, thus inducing structural changes and chemical reactions at room temperature. Consequently, the microstructure of milled powders becomes finer and can be reduced to nanometer range depending on the processing conditions. In certain cases, a process control agent (PCA), which is mostly organic compound, is added to the powder mixture during milling. The PCA adsorbs onto the surface of the powder particles and minimizes cold welding between powder particles, thereby suppressing agglomeration [44]. MA powders are finally consolidated into full density using hot pressing, hipping or extrusion.

The degree of homogenous dispersion of CNTs in metals using MA technique depends on several processing variables such as type of mill, ball-to-powder weight ratio, milling time, process control agent, and so on. Esawi and Morsi investigated the effects of MA time and CNT content on the dispersion of MWNTs in aluminum powder [36, 37]. SEM micrographs showing microstructural evolution of the Al/2 wt% MWNT and Al/5 wt% MWNT powders during mechanical alloying for different periods of time are shown in Figure 2.8. In the process, Al powders (75 μm) and MWNTs (average diameter of ∼140 nm, length of 3–4 μm) of the correct proportion were placed in stainless steel jars containing stainless steel balls with a ball-to-powder weight ratio of 10 : 1. The jars were filled with argon and then agitated using a planetary ball mill at 200 rpm for various milling times. For the Al/2 wt% MWNT sample, aluminum powders were first flattened and formed flakes after milling for 0.5 h due to the high energy collision with stainless steel balls. With increasing milling time to 3 h, the flakes began to weld together, forming large particles of rougher surface. After 6 h, large particles of smooth surface were developed. These particles reached a size of ∼1.6 mm and 3 mm, respectively after 18 h and 24 h of milling. Thus, particle welding is pronounced for Al/2 wt% MWNT sample due to the excellent ductility of Al in this composite and dynamic recovery process during milling. In contrast, large particles are not observed in Al/5 wt% MWNT sample. The incorporation of 5 wt% MWNT into Al impairs its tensile ductility. The large powder sizes obtained for the

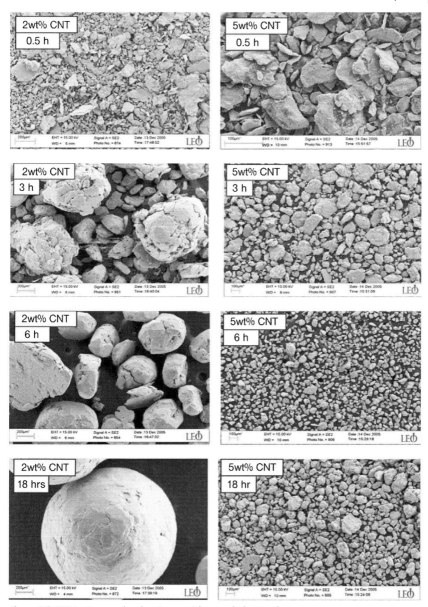

Figure 2.8 SEM micrographs showing particle morphology and size change with mechanically alloying time for Al/2 wt% MWNT and Al/5 wt% MWNT powder composites. Reproduced with permission from [37]. Copyright © (2007) Springer Verlag.

Al/2 wt% MWNT sample can cause serious problems during subsequent sintering or hot pressing. In this case, PCA agent (methanol) is needed to control the size of particles during milling.

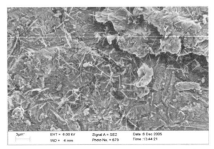

Figure 2.9 CNTs dispersed on the surface of aluminum particles of Al/2 wt% MWNT composite powder after milling for 0.5 h. Reproduced with permission from [36]. Copyright © (2007) Elsevier.

To observe the dispersion of MWNTs in Al matrix, the mill powders were fractured mechanically. Figures 2.9 and 2.10 show SEM fractographs of the Al/2 wt% MWNT sample after milling for 0.5 and 48 h, respectively. The MWNTs are found to disperse uniformly on the flakes' surface after milling for 0.5 h. Carbon nanotubes also remain intact with the aluminum matrix after milling for 48 h. From a series of SEM fractographs examination, MWCNTs are embedded in the aluminum matrix once cold-welding particles of smooth surfaces begin to develop, that is, after milling for 6 h.

More recently, Perez-Bustamante et al. employed MA technique to prepare Al nanocomposites reinforced with 0.25, 0.5, 0.75 and 2 wt% MWNT [38]. Aluminum powders were blended ultrasonically with MWNTs, mechanically milled in a high energy shaker mill for 1–2 h, followed by sintering in vacuum at 550 °C for 3 h. The milling media (hardened steel) to powder ratio was fixed at 5 : 1. Figure 2.11 is the TEM micrograph of as-sintered Al/2 wt% MWNT nanocomposite showing good dispersion of CNTs in aluminum matrix.

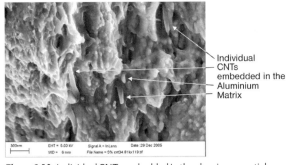

Figure 2.10 Individual CNTs embedded in the aluminum particles of Al/2 wt% MWNT composite powder after milling for 48 h. Reproduced with permission from [36]. Copyright © (2007) Elsevier.

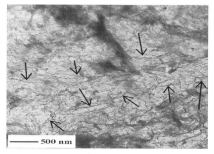

Figure 2.11 Bright field TEM micrograph of MA prepared Al/2 wt % MWNT nano-composite and sintered in vacuum at 550 °C for 3 h. The arrows show the location of MWNTs. Reproduced with permission from [38]. Copyright © (2008) Elsevier.

2.4.3
Controlled Growth of Nanocomposites

He *et al.* developed a novel *in situ* synthesis method for forming CNT(Ni)-Al nanocomposites [46]. The synthesis process involves the initial formation of a Ni(OH)$_2$-Al precursor by means of the chemical deposition-precipitation method. The Ni(OH)$_2$-Al precursor is reduced in hydrogen to yield uniform spreading of Ni nanoparticles on the surfaces of Al powders (Figure 2.12(a)–(c)). Ni-Al powders are placed in a quartz tube reactor where CNTs are synthesized under the flow of mixed CH$_4$/N$_2$/H$_2$ gases at 630 °C. Nickel nanoparticles act as metal catalysts to induce formation of CNTs through decomposition of hydrocarbon gases (Figure 2.12(d)). Subsequently, CNT(Ni)-Al composite powders are cold compressed at 600 MPa, sintered in vacuum at 640 °C for 3 h, and compressed again at 2 GPa to form bulk nanocomposite. Figure 2.13 shows a TEM micrograph of the consolidated Al/5 wt% CNT nanocomposite. The inset reveals that the interfaces of the CNTs and Al are clean, and free from the interfacial reaction products. The synthesized nanocomposite demonstrates improved nanotube dispersion and mechanical properties as a result of uniformly distribution of Ni nanoparticles on the surface of Al powders.

2.4.4
Severe Plastic Deformation

Considerable interest has recently centered on the processing of metallic materials with grain sizes at the nanometer and submicrometer levels. The electronic structure, electrical conductivity, optical and mechanical properties of metals have all been observed to change in the nanoscale region. The mechanical deformation behavior of nanocrystalline metals differs distinctly from that of micrograined metals. As recognized, the yield strength of metals increases significantly by reducing their grain size from micrometer scale to submicrometer or nanometer level. However, nanocrystalline metals exhibit very low tensile ductility due to the lack of strain

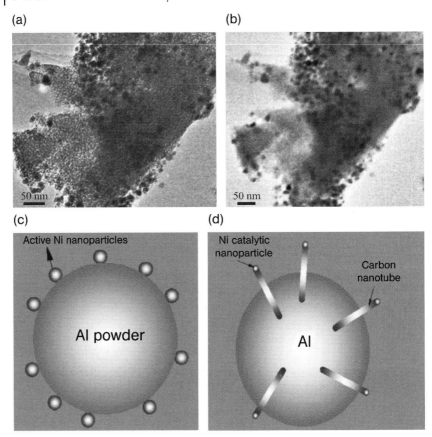

Figure 2.12 Microstructure of the Ni-Al catalyst powders, in which the Ni nanoparticles with a narrow diameter distribution are homogeneously dispersed on the surface of the Al powders. (a) TEM image of a Ni-Al catalyst powder, obtained by reducing the catalyst at 400 °C for 2 h; (b) TEM image of a Ni-Al catalyst powder, showing that the gray Al powder is evenly decorated by several black Ni nanoparticles. Schematic illustrations showing (c) formation of Ni nanoparticles on Al powder by calcinations and reduction of a $Ni(OH)_2$-Al precursor and (d) *in situ* synthesis of CNTs in Al matrix via CVD method. Reproduced with permission from [46]. Copyright © (2007) Wiley-VCH Verlag GmbH.

hardening. Practical application of nanometals is hampered by the difficulties associated with producing them in bulk form. Bulk nanocrystalline metals are generally prepared by consolidation of metal nanopowders using conventional hot pressing, extrusion, and hipping. However, nanograins tend to coarsen into micrometer scale during hot compaction at high temperatures.

Grain refinement of metals to submicrometer range can also be achieved by subjecting the materials to severe plastic deformation such as high pressure torsion (HPT) and equal channel angle pressing (ECAP). In HPT, a disk sample is placed between two anvils under high pressure at room temperature such that it undergoes a very large shear torsion straining. Processing parameters such as the number of

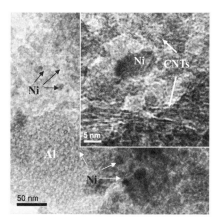

Figure 2.13 TEM image of bulk Al/5 wt% CNT bulk nanocomposite showing homogeneous dispersion of CNTs within the metal matrix. Reproduced with permission from [46]. Copyright © (2007) Wiley-VCH Verlag GmbH.

turns applied to the disk and the magnitude of applied pressure control the final grain sizes of deformed materials. In ECAP, an intense plastic strain is imposed by pressing a sample in a die consisting of two channels equal in cross section, intersecting at an angle ranging from 90 to 157°. The billet retains the same cross-sectional area so that repetitive pressings can be applied to accumulate large plastic strains [47]. In practice, ECAP is more attractive because it can refine the grain sizes of bulk MMCs billets to the submicrometer region [48–50]. ECAP treatment can also improve the homogeneity of reinforcement distribution in the matrix of Al-based MMCs [50]. Apart from grain refinement of metallic materials, HPT has been used to consolidate powder materials to form titanium matrix nanocomposites [51]. An in-depth understanding on the microstructure and deformation behavior of the HPT and ECAP prepared metal-matrix nanocomposites is still lacking.

More recently, Tokunaga *et al.* attempted to use HPT to produce Al-based nanocomposites reinforced with 5 wt% SWNT and 5 wt% fullerene, respectively [52, 53]. In the process, Al powder (75 μm) was mixed with either SWNT or fullerene in ethanol ultrasonically followed by evaporating the solvent. The powder mixture was then placed in a central hole located at the lower anvil. The two anvils were brought into contact to apply a pressure on the disk. The lower anvil was rotated with respect to the upper anvil at a rotation speed of 1 rpm. A pressure of 2.5 GPa was applied during the operation at room temperature. Figure 2.14(a) and (b) show bright field and dark field TEM images for pure Al and Al/5 wt% SWNT nanocomposite, respectively. Large shear straining generates many dislocations that subsequently rearrange into subgrain boundaries. From dark field TEM images, HPT produces finer grain size of Al matrix (∼100 nm) of nanocomposite compared with that of pure aluminum (∼500 nm). The grain size of Al matrix of Al/5 wt% SWNT is even smaller than the grain size reported on the bulk sample deformed by severe plastic deformation. The presence of rings in the selected area diffraction patterns indicates the formation of

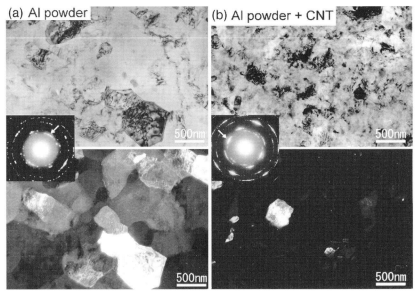

Figure 2.14 Bright field and dark field TEM images as well as selected area diffraction patterns for HPT treated (a) Al powders and (b) mixture of Al powders and CNTs. Reproduced with permission from [52]. Copyright © (2008) Elsevier.

fine crystalline structure. The grain refinement of Al/5 wt% SWNT nanocomposite is due to the presence of CNTs at grain boundaries as they hinder the dislocation annihilation at these sites. Figure 2.15 is HRTEM image showing the presence of a nanotube at a grain boundary. A lattice fringe of CNT wall with a spacing of 0.337 nm at the boundary between two matrix grains A and B can be readily seen. The (1 1 1) lattice fringe with a spacing of 0.234 nm of grain A, and the (2 0 0) lattice fringe with an interval of 0.202 nm of grain B are also observed.

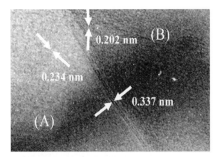

Figure 2.15 Lattice image of HPT treated Al/5 wt% SWNT nanocomposite. Reproduced with permission from [52]. Copyright © (2008) Elsevier.

2.5
Magnesium-Based Nanocomposites

Magnesium is the lightest structural metal with a density of 1.74 g cm^{-3}, which is about two third of density of aluminum (2.70 g cm^{-3}). It also exhibits other advantages such as good mechanical damping properties, good castability (particularly suitable for die casting), and abundant supply globally. Compared with aluminum, magnesium has relatively low strength and ductility, poor creep and corrosion resistance. Magnesium and its alloys are difficult to deform plastically at room temperature due to their hexagonal close-packed (HCP) crystal structure and limited number of slip systems. Despite these deficiencies, magnesium and its alloys have attracted considerable attention in automotive and aerospace industries where light weight is an important consideration. The automotive applications include wheels, steering column lock housings, and manual transmission housings [54]. The low mechanical strength of Mg can be improved by adding ceramic micro- and nano-particles [55–57].

2.5.1
The Liquid Metallurgy Route

2.5.1.1 Compocasting
Very recently, Honma *et al.* fabricated high-strength AZ91D magnesium alloy (Mg-9Al-1Zn-0.3Mn) reinforced with 1.5–7.5 wt% Si-coated carbon nanofibers (CNF) [58]. The CNFs were coated with Si in order to improve the wettability. The nanocomposites were prepared using a compocasting method at semi-solid temperature (585 °C) in a mixed gas atmosphere of SF_6 and CO_2. The resulting nanocomposites were then subjected to squeeze casting and extrusion. The dispersion of CNFs in magnesium alloy matrix is quite uniform due to the improvement in the wettability of CNFs by the Si coating and the break-up of agglomerating clusters by shear deformation during squeeze casting. Compocasting uses semi-solid stirring and is effective for the fabrication of magnesium composites due to its ease of operation and low cost. In the process, reinforcements are introduced into the matrix alloy and stirred in the semi-solid condition. The slurry is then reheated to a temperature above the liquidus of the alloy and poured into the molds. As many pores and segregation are introduced in the composites during compocasting, an additional squeeze casting process is required to minimize the casting defects. This process involves the pouring of molten metal into a pre-heated die and subsequent solidification of the melt under pressure.

2.5.1.2 Disintegrated Melt Deposition
Gupta and coworkers have developed the spraying process termed "disintegrated melt deposition" (DMD) to deposit near net shape metal-matrix microcomposites and nanocomposites [59–61]. The process involves mechanical stirring of molten metal with an impeller under the addition of ceramic particulates, disintegration of the composite slurry by an inert gas jet and subsequent deposition on a substrate in

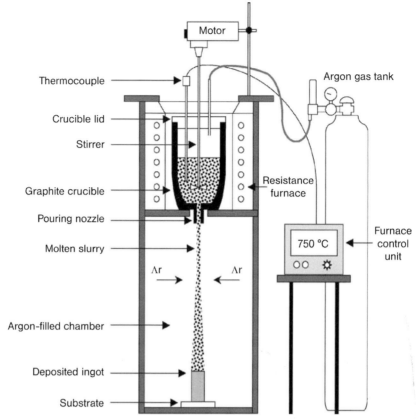

Figure 2.16 Schematic diagram of DMD process. Reproduced with permission from [61]. Copyright © (2004) Elsevier.

the form of a bulk ingot (Figure 2.16). DMD method uses higher superheat temperatures and lower impinging gas jet velocity when compared with conventional spray process. To deposit Mg/MWNT nanocomposites containing 0.3–2 wt% MWNT, magnesium turnings and nanotubes were superheated to 750 °C under an inert argon atmosphere. The molten composite was released through a pouring nozzle located at the base of the crucible and disintegrated using argon gas jets. The ingot was then extruded.

2.5.2
Powder Metallurgy Processing

Gupta and coworkers also prepared the Mg/MWNT nanocomposites containing very low filler loadings (0.06, 0.18 and 0.30 wt%) by conventional PM blending, sintering and hot extrusion [62]. Clustering of MWNTs is observed in the Mg/0.3 wt% MWNT nanocomposite. Thus, PM prepared Mg-based nanocomposites exhibit limited improvement in ultimate tensile strength as expected. Similarly, Carreno-Morelli *et al.* also blended Mg/2 wt% MWNT in a Turbula mixer, followed by hot pressing and hot isostatic pressing [19]. Very little improvement in mechanical strength is found for such nanocomposites.

Reducing the length of CVD grown CNTs is found to be very effective to fully incorporate and disperse CNTs into composites. The length of CVD-grown CNTs can be controlled mechanically through ultrasonication, ball milling, and high speed shearing. Very recently, Shimizu *et al.* used a high-speed blade cutting machine to reduce the length of CVD prepared MWNTs [63]. Damaged MWNTs with an average length of ~5 µm were mixed mechanically with AZ91D magnesium alloy powders in a mill containing zirconia balls under a protective argon atmosphere. Milled composite powder mixtures were then hot pressed and extruded into rods (Figure 2.17(a)–(e)). At 1 wt% MWNTs, CNTs are distributed uniformly in magnesium powders (Figure 2.17(c)). When the filler content is increased to 5 wt%, agglomeration of MWNTs occurs (inset of Figure 2.17(d)).

Figure 2.17 SEM images of (a) mechanically milled AZ91D magnesium alloy powders of ~100 µm; (b) shortened CNTs with an average length of ~5 µm; (c) mechanically mixed AZ91D/1 wt% CNT powders; (d) mechanically mixed AZ91D/5 wt% CNT powders; (e) extruded AZ91D/CNT nanocomposite rods. Reproduced with permission from [63]. Copyright © (2008) Elsevier.

2.5.3
Friction Stir Processing

Friction stir processing (FSP) is a solid state processing method, developed on the basic principles of friction stir welding; it is an effective tool to form fine-grained microstructures in near surface region of metallic materials. In the process, a rotating pin tool is inserted to the substrate, the resulting frictional heating and the large processing strain produce microstructural refinement, densification and homogenization [64]. Mishra et al. fabricated the Al-based surface composites by dispersing SiC powders (0.7 μm) on the surface of Al 5083 Al (Al-Mg-Mn) alloy plate followed by FSP treatment [65]. They reported that SiC microparticles were well bonded and dispersed in the Al matrix of surface composite layer ranged from 50 to 200 μm. Furthermore, FSP resulted in significant grain refinement in surface layer.

Recently, Morisada et al. fabricated AZ31/MWCNT surface nanocomposite by using the FSP technique [66]. MWCNTs were initially filled into a groove on the AZ31 (Mg-3Al-1Zn) alloy plate followed by the insertion of a pin into the groove. The tool was rotated at a fixed speed of 1500 rpm, but its travel speed varied from 25 to 100 mm min^{-1}. The dispersion of CNTs depend greatly on the travel speed of pin tool. Figure 2.18(a)–(f) show SEM micrographs of AZ31/MWCNT surface nanocomposite specimens prepared from FSP with the pin traveling at different speeds. At a high travel speed of 100 mm min^{-1}, entangled CNTs in the form of large clusters can be readily seen in the surface composite (Figure 2.18(b)). This is because the high travel speed cannot produce sufficient frictional heat to mix CNTs properly in the alloy (Figure 2.18(d). Large MWNT clusters tend to break up into smaller agglomerates by decreasing the travel speed of pin tool. A better distribution of MWNTs is achieved at a low speed of 25 mm min^{-1}(Figure 2.18(f)). Careful examination of this micrograph reveals that CNTs are still agglomerated into small clusters. Much more work is needed in future to improve the dispersion of nanotubes using FSP.

2.6
Titanium-Based Nanocomposites

Titanium alloys are structural materials widely used in aerospace, chemical, biomedical and automotive industries due to their excellent mechanical properties at room temperature. However, titanium alloys exhibit low mechanical strength at high temperatures. Continuous SiC fibers and TiC particulates have been incorporated into Ti alloys to enhance their mechanical performances [67–71].

Very little information is available in the literature regarding the fabrication and microstructural behavior of Ti-based nanocomposites [72]. The difficulty in dispersing CNTs in high reactivity titanium matrix hinders the development of Ti/CNT nanocomposites. This is a big challenge for materials scientists: to produce Ti/CNT nanocomposites with desired mechanical properties using novel processing techniques.

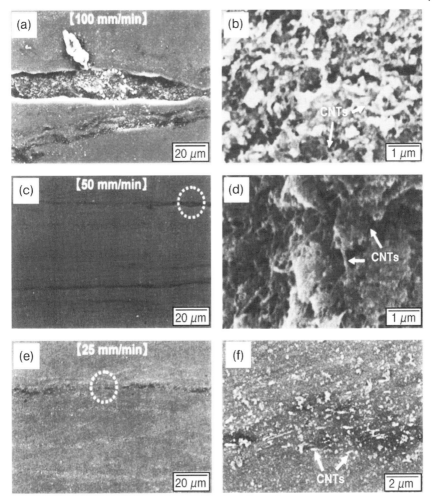

Figure 2.18 SEM micrographs of the AZ31/MWCNT surface nanocomposite specimens prepared by the FSP technique. (b), (d) and (f) are higher magnification views inside a dotted circle of (a), (c) and (e), respectively. Reproduced with permission from [66]. Copyright © (2006) Elsevier.

2.7
Copper-Based Nanocomposites

Copper is a ductile metal having low mechanical strength but excellent electrical and thermal conductivities. To enhance the mechanical strength, ceramic reinforcements are added to copper [73–76]. In recent years, microelectonic devices made from integrated circuits generate enormous local heat during operation. There is a growing demand for high strength Cu-based composite materials in microelectronic packaging. However, poor compatibility between SiC and Cu is the main problem in the fabrication of Cu/SiC composites. Therefore, SiC whiskers and microparticulates as well as SiC nanoparticles are coated with copper prior to the composite fabrication [77–79]. Furthermore, a large volume content of ceramic reinforcement is needed in the composite materials for thermal management applications. The CNTs and CNFs that have excellent thermal conductivity are ideal reinforcing filler materials for Cu-based composites for thermal management applications. Only low loading levels of CNTs are added in the composites to achieve these purposes.

2.7.1
Liquid Infiltration

Jang et al. employed the liquid infiltration process to incorporate VGCNFs into copper [80]. Entangled nanofibers were filled into a copper tube and mechanically drawn in order to form straight fibers. Bundle of drawn Cu tubes were placed in a mold, heated in a furnace at 1100 °C for 10 min, and transferred to a press for compaction under the pressure of 50 MPa. The volume content of VGCNF in the resulting composite was 13%. SEM examination revealed that the VGCNFs were not uniformly distributed throughout the composite due to the flow of molten metal during melting process.

2.7.2
Mechanical Alloying

Hong and coworkers fabricated the Cu/MWNT nanocomposites using MA technique [81]. Copper nanopowders (200–300 nm) were prepared by spray drying of [Cu$(NO_3)_2$].$3H_2O$ water solution on hot walls. The resulting powders were heat treated at 300 °C followed by reduction in hydrogen atmosphere. CVD-grown MWNTs with an average diameter of 40 nm were mixed with Cu nanopowders in a high energy planetary mill for 24 h. The milled powders were cold compacted and spark plasma sintered at 700 °C. The sintered Cu/MWCNT nanocomposites were cold rolled up to 50% reduction and followed by annealing at 650 °C for 3 h. Both Cu/5 vol% MWNT and Cu/10 vol% MWNT nanocomposites were fabricated.

Figure 2.19(a) shows an SEM micrograph of the mechanically milled Cu/MWNT composite powder. Copper nanopowders experience cold welding and fracturing during mechanical milling, leading to the formation of welded powders with sizes of ∼10 µm. High magnification SEM micrography reveals that the CNTs are distributed on the surface of a Cu powder (Figure 2.19(b)). The surfaces of Cu powders where MWNTs are located result in the formation of Cu/MWNT composite during spark plasma sintering. The inner region of cold-welded Cu powders with few or no CNTs yields MWNT-free copper matrix region during sintering (Figure 2.20(a) and (b)). Cold rolling of consolidated composite products has resulted in the alignment of MWNTs along the rolling direction. In general, homogeneous distribution of CNTs in metals through mechanical alloying depends greatly on the milling time, ball-to-powder ratio, milling atmosphere and

2.7 Copper-Based Nanocomposites

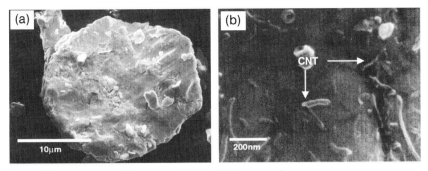

Figure 2.19 (a) Low and (b) high magnification SEM images of mechanically milled Cu/MWNT powder. Reproduced with permission from [81]. Copyright © (2006) Elsevier.

so on. It is necessary to optimize milling conditions to obtain uniform dispersion of CNTs throughout entire copper matrix.

Spark plasma sintering (SPS) is an effective tool to sinter and consolidate powders under the application of an external pressure at relatively low temperatures and short periods of time (Figure 2.21). It is particularly useful for sintering nanopowders because the grain coarsening problem can be minimized or avoided. Retention of nanograins enables the resulting composites exhibiting enhanced mechanical strength and hardness. SPS is a pressure assisted sintering technique in which a pulsed current is applied to the upper and lower graphite plungers such that high temperature plasma is generated between the gaps of electrodes. In this case, uniform heating can be achieved for sintering compacted powder specimen. Sintering can be performed at wide range of pressures and temperatures. The technique is widely used to prepare densely structural ceramics, nanoceramics and their composites as well as bulk CNTs [82–84].

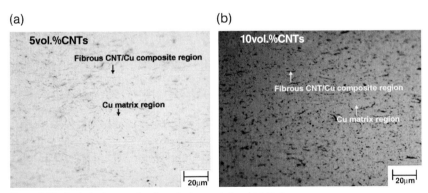

Figure 2.20 Optical micrographs of spark plasma sintered (a) Cu/5 vol% MWNT and (b) Cu/10 vol% MWNT nanocomposites. Reproduced with permission from [81]. Copyright © (2006) Elsevier.

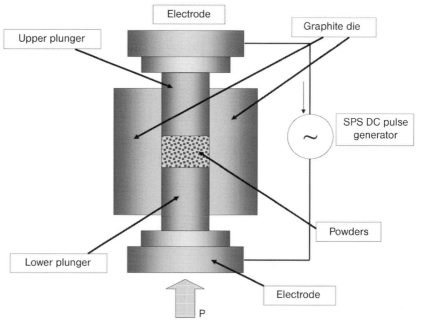

Figure 2.21 Schematic representation of spark plasma sintering system. Reproduced with permission from [83]. Copyright © (2007) Elsevier.

2.7.3
Molecular Level Mixing

An important step procedure for forming metal nanocomposites at molecular level is functionalization of CNTs. This strategy is often employed to prepare polymer-CNT nanocomposites since CNTs have poor compatibility with most polymers [6, 85]. CNTs can be functionalized covalently by oxidizing nanotubes in nitric and sulfuric acids in order to induce carboxylic or hydroxyl groups on the end-caps or defect sites of CNTs. These acids disrupt the aromatic ring arrangement at the caps of CNTs, leading to the incorporation of functional groups at the open ends. This acid treatment also results in shortening of CNTs as described in Chapter 1. To form nanocomposites, functionalized CNTs act as adsorption centers that strongly interact with metal ions or hydration molecules in aqueous phase. In other words, functionalized CNTs react with metal ions of selected metal salt and organic ligand in a solution at the molecular level to form metal complexes which act as the building blocks for metal-matrix nanocomposites.

Hong and coworkers synthesized the Cu/MWNT nanocomposites using a molecular level mixing process [86–89]. The synthesis involves several steps: (a) functionalization of MWNTs and their dispersion in ethanol; (b) addition of copper acetate monohydrate [$Cu(CH_3COO)_2 \cdot H_2O$] to CNT suspension under sonication; (c) dissolution of copper salt and attachment of Cu ions to the functional

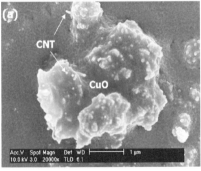 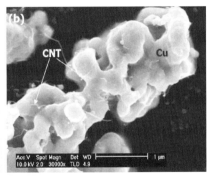

Figure 2.22 SEM micrographs of (a) CuO/MWNT composite powders after calcination and (b) Cu/MWNT composite powders after reduction in hydrogen atmosphere. Reproduced with permission from [87]. Copyright © (2007) Elsevier.

group of MWNTs; (d) vaporization of the solvent and calcination of vaporized powders in air to form CuO/MWNT powder; (e) reduction of CuO/MWNT powder in hydrogen atmosphere to form Cu/MWNT composite powder and (f) spark plasma sintering at 550 °C for 1 min under 50 MPa.

Figure 2.22(a) and (b) show the morphologies of CuO/MWNT and Cu/MWNT composite powders, respectively. The step processes (a)–(d) above produce CuO powders of several micrometers. The CuO of composite powder is then reduced to Cu under hydrogen atmosphere. Molecular level mixing enables nanotubes to be implanted within Cu powders rather than attached on the surface of Cu powders. The appearance of spark plasma sintered nanocomposites are shown in Figure 2.23(a). To observe the dispersion of nanotubes, the consolidated composites were chemically etched. The microstructure of etched Cu/5 vol% MWNT composite shows homogeneous dispersion of MWNTs within the copper matrix (Figure 2.23(b)). The cracks in the micrographs originate from the chemical etching.

From these, it appears that two kinds of oxygen atoms exist in the CuO/MWNT powder: one is associated with functional carboxyl group on nanotubes and the other derives from calcination of Cu atoms. Oxygen atoms of the CuO matrix can be reduced by hydrogen gas while those in carboxyl group of nanotubes cannot be easily reduced (Figure 2.24). Residual oxygen at the interface enhances the bonding between nanotubes and Cu, thereby yielding homogeneous dispersion of nanotubes in the Cu matrix [89]. Such strong interfacial bonding facilitates effective load transfer from the matrix to nanotubes during mechanical loading. Figure 2.25(a) is a higher TEM image showing the copper matrix (denoted as A), MWNT/Cu interface (B) and CNT (C) of the Cu/10 vol% MWNT nanocomposite. The energy-dispersive X-ray (EDX) spectra of these areas clearly demonstrate that the MWNT-Cu interface is enriched with oxygen (Figure 2.25(b)). A similar finding is also observed in Fourier-transform infrared (FTIR) spectra of the Cu/10 vol% MWNT composite powder (Figure 2.25(c)). The Cu—O stretching bond at 630 cm^{-1} can still be observed in

(a)

(b)

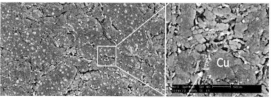

Figure 2.23 (a) Appearances of spark plasma sintered copper, Cu/5 vol% MWNT and Cu/10 vol% MWNT nanocomposites; (b) SEM image of etched Cu/5 vol% MWNT nanocomposite showing homogeneous distribution of CNTs within copper matrix. Reproduced with permission from [87]. Copyright © (2007) Elsevier.

the Cu/10 vol% MWNT composite powder even after reduction of CuO matrix in hydrogen.

Xu et al. also fabricated Cu/MWNT spherical powders by mixing Cu ions with functionalized MWNTs at the molecular level [90]. They managed to produce Cu/MWNT composite powders having much finer sizes of ∼200–300 nm. The MWNTs were dispersed ultrasonically in a gelatin solution followed by adding

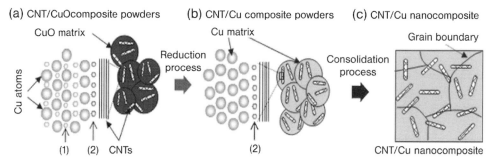

(1) Oxygen atoms in the CuO matrix formed by oxidation process
(2) Oxygen atoms stemming from functional groups on the CNT surface.

Figure 2.24 Schematic diagrams showing (a) oxygen atoms in CuO matrix and functionalized MWNTs; (b) residual oxygen atoms at the nanotube-Cu interface after reduction process; (c) homogeneous dispersion of nanotubes in Cu matrix after consolidation. Reproduced with permission from [89]. Copyright © (2008) Wiley-VCH Verlag GmbH.

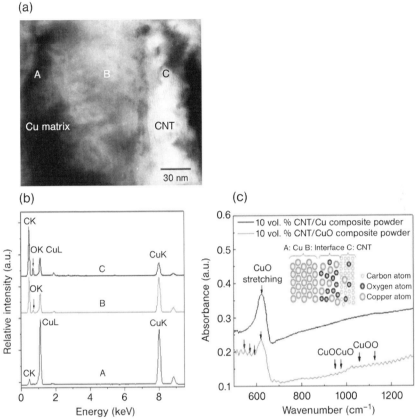

Figure 2.25 (a) TEM micrograph showing Cu matrix, MWNT-Cu interface and nanotube of Cu/10 vol% MWNT nanocomposite; (b) EDX spectra of Cu matrix, MWNT-Cu interface and nanotube; (c) FTIR spectra of CuO/10 vol% MWNT and Cu/10 vol% MWNT composite powders. Reproduced with permission from [89]. Copyright © (2008) Wiley-VCH Verlag GmbH.

copper sulfate and glucose. The gelatine with amine groups wraps the MWNTs owing to the electrostatic attraction. Moreover, the gelatin tends to attract Cu^{2+} ions, forming a copper complex that reacts with hydroxyl in the presence of glucose. At this stage, Cu^{2+} is reduced to Cu^{+} and small CuO particles nucleate on the surface of MWNTs. Final reduction in hydrogen atmosphere yields Cu/MWNT nanocomposite powders.

2.7.4
Electrodeposition

Among various processing techniques for making metal-matrix nanocomposites, electrodeposition has advantages for forming dense composite materials. These include low cost, ease of operation, versatility and high yield. Both direct current (d.c.)

and pulse deposition (plating) have been used extensively for the electrodeposition of metals/alloys. Pulse plating offers more uniform deposition of coating than the d.c. method. Several parameters such as peak current density, frequency and duty cycle affect the adsorption and desorption of chemical species in the electrolyte. The electrodeposition parameters can be tailored to produce deposits of desired grain size, microstructure and chemistry. Therefore, electrodeposition has emerged as an economically viable process to produce pore-free nanostructured metals and nanocomposites in the form of coatings and thick plates [91–96]. Nanocrystalline coatings can be deposited on the cathode electrode surface by properly monitoring the electrodeposition conditions such as bath composition, temperature, pH, additive agent and deposition time.

Very recently, Chai et al. prepared Cu/MWNT nanocomposite by means of direct current electrochemical co-deposition technique [97]. The MWNTs were suspended in the copper plating solution initially. Both nanotubes and copper ions were driven towards the cathode and deposited onto the cathode simultaneously during the deposition. Homogenous dispersion of MWNTs in copper matrix was observed.

As CNTs have poor compatibility with copper, special surface treatments are needed to modify the surfaces of CNTs to enhance interfacial interaction. Lim et al. used electrodeless plating to deposit a nickel layer on the surface of SWNTs fabricated by a HiPCo method [98]. They reported that the interfacial bonding was significantly improved after coating the nanotubes with a layer of nickel by eletrodeless plating. The coated nanotubes were then blended with copper powders by means of mechanical mixing process to form the Cu/SWNT nanocomposites.

2.7.5
Patent Process

Copper-based nanocomposites filled with low loading levels of CNTs having excellent thermal conductivity are potential composite materials for thermal management applications in electronic industries. As mentioned above, substantial progress has been made in worldwide research laboratories in the development, processing and characterization of Cu/CNT nanocomposites. Very recently, Hong et al. have invented a technical process for possible commercialization of Cu/MWNT nanocomposite powders [99]. Their invention discloses a process of producing a metal nanocomposite powder homogeneously reinforced with CNTs. The invention consists of an initial dispersion of MWNTs in a predetermined dispersing solvent such as ethanol under sonication followed by the copper salt ($Cu(CH_3COO)_2$) addition. The resulting dispersion solution is again subjected to sonication for 2 h to ensure even dispersion of the CNT and copper molecules in the solution and to induce the chemical bond between molecules of the CNT and copper. This mixed solution is finally heated at 80–100 °C to vaporize water, and calcined at 350 °C to form stable CuO/MWNT powders. Such composite powders are reduced at 200 °C under a hydrogen gas atmosphere to form the Cu/MWNT powders.

2.8
Transition Metal-Based Nanocomposites

2.8.1
Ni-Based Nanocomposites

Electrodeposited Ni and its nanocomposites reinforced with ceramic nanoparticles have been studied extensively. Electrodeposited Ni coatings find important application as bulk nanostructured materials for mechanical characterization [91–93]. This is because bulk metals and composites prepared by direct hot compaction of nanopowders undergo excessive grain growth, leading to poor mechanical properties in tensile measurements. Electrodeposited Ni coatings with nanograins generally exhibit much higher tensile strength and stiffness but poorer tensile ductility compared with their micrograin counterparts. It is considered that CNTs with superior tensile strength, stiffness and fracture strain can further enhance the mechanical performances of Ni coatings.

In the deposition of Ni coatings, bath composition and plating condition (d.c. or pulse) play a crucial role on the morphology and dispersion of CNTs in Ni coatings (Table 2.5). Oh and coworkers prepared Ni/MWNT nanocomposite coatings by using d.c. plating [100]. The electrolyte used was typical sulfate Watts bath. To assist the dispersion of CVD grown MWNTs, sodium dodecyl sulfate (SDS) and hydroxypropylcellulose (HPC) were added into the electrolyte. The total amount of SDS and HPC was fixed at $10\,g\,L^{-1}$. The length of CVD-grown MWNTs was reduced from ~20 μm to less than 5 μm by milling with zirconia balls for 24 h. The thickness of electrodeposited coatings was controlled to 50 μm.

Figure 2.26(a)–(e) show field-emission SEM micrographs of Ni/CNT nanocomposite deposited in a bath containing different additive contents. The MWNT concentration of the electrolyte is maintained at $10\,g\,L^{-1}$, corresponding to 14.6 vol% CNT. From Figure 2.26(a) and (e), it is evident that SDS is more effective for CNT dispersion than HPC. Furthermore, the incorporation of MWNTs into Ni matrix is enhanced by adding SDS-HPC mixture to the electrolyte (Figure 2.26(b) and (c)). Using the same technique, Oh and coworkers also prepared the Sn/MWNT lead-free solder for electronic packaging applications [101].

Table 2.5 Bath composition for deposition of Ni/CNT coatings.

Plating condition	Bath Composition ($g\,L^{-1}$)						Temperature (°C)	Bath pH
	$NiSO_4$	$NiCl_2$	H_3BO_3	Saccharin	SDS	HPC		
d.c. Ref [99]	260	45	15	0.5	2.5–10	2.5–10	40	—
d.c. Ref [102]	315	25	35	0.1	0.1	—	60	3.5
Pulse-reverse Ref [104]	315	25	35	0.1	0.1	—	60	3.5
Pulse-reverse Ref [105]	280	35	45	—	—	—	54	4

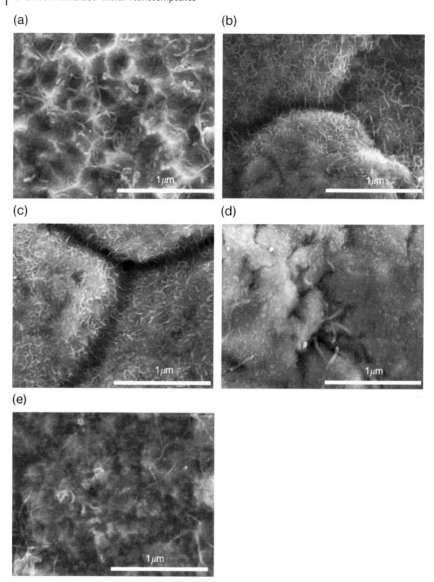

Figure 2.26 Field emission SEM micrographs of the Ni/MWNT nanocomposite electrodeposited from a bath containing dispersion additive of (a) 10 g L^{-1} SDS; (b) 7.5 g L^{-1} SDS and 2.5 g L^{-1} HPC; (c) 5 g L^{-1} SDS and 5 g L^{-1} HPC; (d) 2.5 g L^{-1} SDS and 7.5 g L^{-1} HPC; (e) 10 g L^{-1} HPC. Reproduced with permission from [100]. Copyright © (2008) Elsevier.

Dai *et al.* prepared the Ni/MWNT deposit from a modified bath solution at 60 °C [102]. CNTs are homogeneously dispersed within the nickel matrix (Figure 2.27). Dark-field TEM image reveals that the nickel matrix grains exhibit

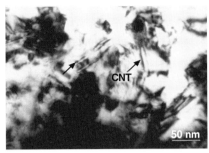

Figure 2.27 TEM image of electrodeposited Ni/MWNT nanocomposite showing uniform distribution of CNTs. Reproduced with permission from [102]. Copyright © (2008) Elsevier.

a mean grain size of 28 nm (Figure 2.28). The formation of nanograins coupled with CNT reinforcement lead to the enhancement of mechanical properties as expected.

Compared with d.c. deposition, pulse-plating and pulse-reverse plating facilitate larger amounts of filler incorporation into the matrix of metal-matrix nanocomposites under the same current density [103, 104]. For example, the maximum alumina incorporation in nickel and copper matrix using direct current method is ~1.5 wt% and 3.5 wt%, respectively. However, a maximum incorporation of 5.6 wt% alumina in a copper matrix can be achieved by pulse plating [103]. Sun *et al.* used deposited Ni/SWNT and Ni/MWNT nanocomposite coatings in the Watt bath solution using d.c. and pulse-reverse plating methods [105]. As expected, the d.c. method produces uneven surface coating with many deposit protrusions. The coatings formed by the pulse-reverse method exhibit good interfacial bonding between CNTs and the metal matrix. Guo *et al.* demonstrated that higher pulse frequency and higher pulse reverse ratio employed during electrodeposition of

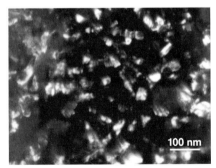

Figure 2.28 TEM dark-field image showing formation of Ni nanograins (28 nm) in Ni/MWNT composite prepared by electrodeposition. Reproduced with permission from [102]. Copyright © (2008) Elsevier.

Ni/MWNT coatings in a Watt-type bath producing more homogeneous or smooth surface coatings [106].

It is widely recognized that Ni and Ni-P metallic coatings can also be deposited onto a substrate by means of electrodeless plating. Electrodeless Ni-P coatings have been used in many industrial applications due to their good corrosion and wear resistance as well as uniformity of thickness. Electrodeless plating can produce very smooth metallic coating from a solution via several chemical reactions without using electric power. Other advantages include ease of operation and good coverage in blind holes. Therefore, this technique is also very effective to deposit dense and uniformly distributed nanotubes in the matrices of CNT–metal nanocomposites [107–109].

2.8.2
Co-Based Nanocomposites

Hong and coworkers prepared *in situ* Co/MWNT nanocomposites by using a molecular level mixing method [110]. In the process, functionalized MWNTs were dispersed in oleylamine ultrasonically followed by adding the Co(II) acetylacetonate (Co(acac)$_2$). The copper salt decomposed in oleylamine during refluxing in argon atmosphere into cube-shaped CoO nanoparticles of $\sim$50 nm. For comparison, pristine CoO nanopowders were also fabricated under similar conditions (Figure 2.29(a) and (b)). The X-ray diffraction pattern as shown in Figure 2.29(d)) confirms the formation of CoO nanocrystals in nanocomposite powder. The size of CoO nanocrystals is much smaller than that of CuO/MWNT composite powder (several micrometers) as described above. The CoO/MWNT powder was reduced in hydrogen atmosphere to form the Co/MWNT composite powder. Such composite powder was consolidated by spark plasma sintering to produce bulk Co/7% MWNT nanocomposite. The grain size of consolidated Co/MWNT nanocomposite is $\sim$310 nm and smaller than that of pure Co of $\sim$350 nm (Figure 2.30(a) and (b)). The SEM and high-resolution TEM micrographs reveal homogeneous dispersion of MWNTs within the Co matrix (Figure 2.30(c) and (d)).

As functionalization causes structural damage to nanotubes, Hong and coworkers used pristine MWNTs instead to prepare *in situ* Co/MWNT nanocomposites [111]. They dispersed pristine nanotubes in dioctyl ether (solvent) with oleylamine as a surfactant under mechanical stirring (Figure 2.31(a)). Subsequently, Co(acac)$_2$ and 1,2-hexadecanediol (reducing agent) were mixed with the nanotube suspension, followed by heating to the refluxing temperature of solvent. The nanotubes dispersed homogeneously in the solvent are effective nucleation sites for the heterogeneous nucleation of Co atoms (Figure 2.31(b)). The Co nuclei then grew and coalesced to form pearl-necklace-structured Co/MWNT powders in which Co nanoparticles are threaded by nanotubes (Figure 2.31(c)–(e)). No calcination or oxidation of matrix material is needed. Dense Co/MWNT nanocomposite can be prepared by directly sintering the synthesized powders. This procedure yields weaker interfacial nanotube-matrix bonding due to the absence of interfacial oxygen atoms. However, other functional properties, such as field emission, are enhanced

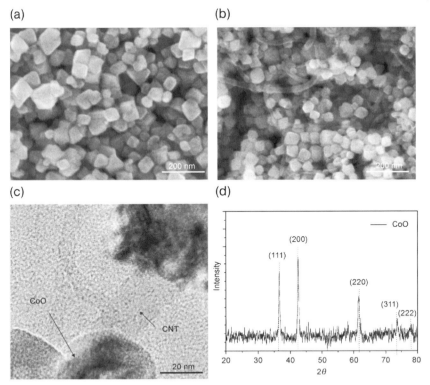

Figure 2.29 SEM micrographs of (a) CoO nanocrystals and (b) CoO/MWNT nanopowders containing 7 vol.% MWNT; (c) high-resolution TEM image of the area near the MWNT-CoO interface; (d) X-ray diffraction pattern of the CoO/MWNT nanopowders. Reproduced with permission from [110]. Copyright © (2007) Wiley-VCH Verlag GmbH.

due to the absence of oxygen transient bonding layer. As mentioned above, residual oxygen is found at the nanotube-matrix interface by using funtionalized nanotubes as precursor materials. Residual oxygen enhances interfacial bonding and mechanical strength of the composites, but degrades their electrical and thermal properties. This is because interfacial oxygen bonding layers act as a scattering center for electron and phonon.

2.8.3
Fe-Based Nanocomposites

All the fabrication techniques for making metal–CNT nanocomposites described above can be referred to as "*ex situ*" since the CNTs have been synthesized independently and then introduced directly into metal matrices of composites during processing. On the other hand, *in situ* CNT reinforcement can be synthesized *in situ* in the metal matrix using transition metal catalysts through the CO

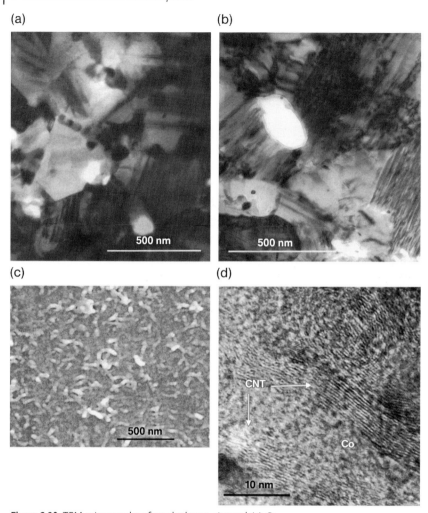

Figure 2.30 TEM micrographs of spark plasma sintered (a) Co and (b) Co/7 vol% MWNT nanocomposite with ultrafine-grained matrix. (c) SEM and (d) high-resolution TEM micrographs of Co/7 vol% MWNT nanocomposite. Reproduced with permission from [110]. Copyright © (2007) Wiley-VCH Verlag GmbH.

disproportionation method. Very recently, Goyal et al. prepared *in situ* Fe/SWNT and Fe/MWNT nanocomposites using CO disproportionation method [112, 113]. To prepare the Fe/MWNT nanocomposite, the Co catalyst and catalyst precursors, that is iron acetate (0.01 wt%) and cobalt acetate (0.01 wt%), were dissolved in ethanol. Iron powder was soaked in this solution, dried and pressed into pellets. The porosities of pellets were controlled by varying the pelletizing pressure. The pellets were placed in a quartz boat located inside a horizontal quartz tube reactor in a high temperature furnace (800 °C). Mixed acetylene, CO and argon

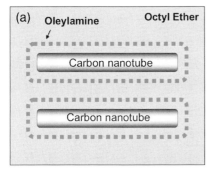
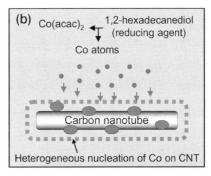
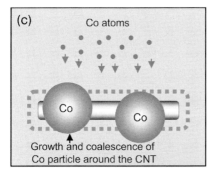
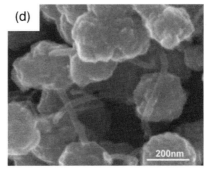
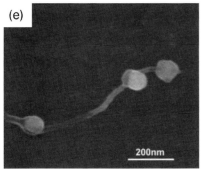

Figure 2.31 Schematic representation of the formation of Co/MWNT powders: (a) Dispersion of MWNTs in dioctyl ether and oleylamine; (b) heterogeneous nucleation of Co nuclei on MWNTs' (c) growth and coalescence of Co nuclei to form composite powders; (d) and (e) SEM micrographs of composite powders showing formation of pearl-necklace like structure. Reproduced with permission from [111]. Copyright © (2006) Wiley-VCH Verlag GmbH.

gases were introduced into the reactor at atmospheric pressure [113]. Figure 2.32 (a)–(d) show SEM micrographs of the Fe/MWNT nanocomposite. The micrographs reveal dense growth of MWNTs in the matrix and within cavities of the composite. This is because iron also serves as an excellent catalyst for nanotube growth. The CNTs form bridges across the cavities in the iron matrix.

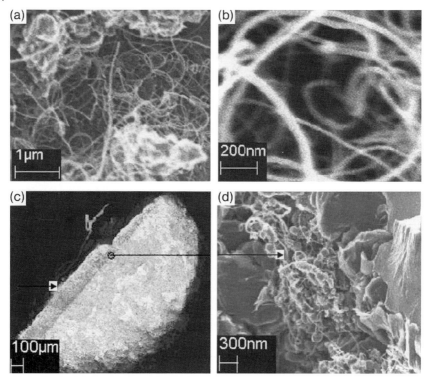

Figure 2.32 SEM micrographs showing formation of *in situ* CNTs in Fe/MWNT nanocomposite. (a) and (c) are low magnification images. (b) and (d) are higher magnification images showing MWNTs in a dense network within cavities and matrix of the composite. Reproduced with permission from [113]. Copyright © (2007) Elsevier.

References

1 Tjong, S.C. and Ma, Z.Y. (2000) Microstructural and mechanical characteristics of *in-situ* metal matrix composites. *Materials Science and Engineering R*, 29, 49–113.
2 Walsch, P.J. (2001) Carbon Fibers, in *Materials Handbook*, vol. 21, ASM International, Materials Park, Ohio, USA, DOI:10.1361/asmhb0003354.
3 http://www.cytec.com.
4 Matsunaga, T., Matsuda, K., Hatayama, T., Shinozaki, K. and Yoshida, M. (2007) Fabrication of continuous carbon-fiber-reinforced aluminum-magnesium alloy composite wires using ultrasonic filtration method. *Composites A*, 38, 1902–1911.
5 Goh, C.S., Wei, J., Lee, L.C. and Gupta, M. (2006) Simultaneous enhancement in strength and ductility by reinforcing magnesium with carbon nanotubes. *Materials Science and Engineering A*, 423, 153–156.
6 Zhang, Q., Wu, G., Chen, G., Jiang, L. and Luan, B. (2003) The thermal expansion and mechanical properties of high

reinforcement content SiC$_p$/Al composites fabricated by squeeze casting technology. *Composites A*, **34**, 1023–1027.

7 Nam, T.H., Requena, G. and Degischer, P. (2008) Thermal expansion behavior of aluminum matrix composites with densely packed SiC particles. *Composites A*, **39**, 856–865.

8 Ma, Z.Y. and Tjong, S.C. (2000) High temperature creep behavior of SiC particulate reinforced Al-Fe-V-Si alloy composite. *Materials Science and Engineering A*, **278**, 5–13.

9 Tjong, S.C. and Ma, Z.Y. (1999) High-temperature creep behavior of powder-metallurgy aluminum composites reinforced with SiC particles of various sizes. *Composites Science and Technology*, **59**, 1117–1125.

10 Lewis, C.A. and Withers, P.J. (1995) Weibull modeling of particle cracking in metal matrix composites. *Acta Metallurgica*, **43**, 3685–3699.

11 Spowart, J.E. and Micracle, D. (2003) The influence of reinforcement morphology on the tensile response of 6061/SiC/25p discontinuously-reinforced aluminum. *Materials Science and Engineering A*, **357**, 111–123.

12 Shen, Y.L., Williams, J.J., Piotrowski, G., Chawla, N. and Guo, Y.L. (2001) Correlation between tensile and indentation behavior of particle-reinforced metal matrix composites: An experimental and numerical study. *Acta Materialia*, **49**, 3219–3229.

13 Ma, Z.Y., Li, Y.L., Liang, Y., Zheng, F., Bi, J. and Tjong, S.C. (1996) Nanonometric Si$_3$N$_4$ particulate-reinforced aluminum composite. *Materials Science and Engineering A*, **219**, 229–231.

14 Tun, K.S. and Gupta, M. (2007) Improving mechanical properties of magnesium using nano-yttria reinforcement and microwave assisted powder metallurgy method. *Composites Science and Technology*, **67**, 2657–2664.

15 Cao, G. and Li, X. (2008) Mechanical properties and microstructure of SiC- reinforced Mg-(2,4)Al-Si nanocomposites fabricated by ultrasonic cavitation based solidification processing. *Materials Science and Engineering A*, **486**, 357–362.

16 Kuzumaki, T., Miyazawa, K., Ichinose, H. and Ito, K. (1998) Processing of carbon nanotube reinforced aluminum composite. *Journal of Materials Research*, **13**, 2445–2449.

17 Ci, L., Ryu, Z., Phillipp, N.Y. and Ruhle, M. (2006) Investigation of the interfacial reaction between multi-walled carbon nanotubes and aluminum. *Acta Materialia*, **54**, 5367–5375.

18 Chen, W.X., Tu, J.P., Wang, L.Y., Gan, H.Y., Xu, Z.D. and Zhang, X.B. (2003) Tribological application of carbon nanotubes in a metal-based composite coating and composites. *Carbon*, **41**, 215–222.

19 Carreno-Morelli, E., Yang, J., Couteau, E., Hernadi, K., Seo, J.W., Bonjour, C., Forro, L. and Schaller, R. (2004) Carbon nanotube/magnesium composites. *Physica Status Solidi A-Applied Research*, **201**, R53–R55.

20 Lau, K.T. and Hui, D. (2002) Effectiveness of using carbon nanotubes as nano-reinforcements for advanced composite structures. *Carbon*, **40**, 1597–1617.

21 Viswanathan, V., Laha, T., Balani, K., Agarwal, A. and Seal, S. (2006) Challenges and advances in nanocomposite processing techniques. *Materials Science and Engineering R*, **54**, 121–285.

22 Tjong, S.C., Ku, J.S. and Wu, C.S. (1994) Corrosion behavior of laser consolidated chromium and molybdenum plasma spray coating on Fe-28Mn-7Al-1C alloy. *Scripta Metallurgica et Materialia*, **31**, 835–839.

23 Laha, T., Agarwarl, A., McKechnie, T. and Seal, S. (2004) Synthesis and characterization of plasma spray formed carbon nanotubes reinforced aluminum composite. *Materials Science and Engineering A*, **381**, 249–258.

24 Laha, T., Kuchibhatla, S., Seal, S., Li, W. and Agarwal, A. (2007) Interfacial

phenomena in thermally sprayed multiwalled carbon nanotube reinforced aluminum nanocomposite. *Acta Materialia*, **55**, 1059–1066.

25 Laha, T., Liu, Y. and Agarwarl, A. (2007) Carbon nanotubes reinforced aluminum nanocomposite via plasma and high velocity oxy-fuel spray forming. *Journal of Nanoscience and Nanotechnology*, **7**, 515–524.

26 Laha, T. and Agarwarl, A. (2008) Effect of sintering on thermally sprayed carbon nanotube reinforced aluminum nanocomposite. *Materials Science and Engineering A*, **480**, 323–332.

27 Zhou, S.M., Zhang, X.B., Ding, Z.P., Min, C.Y., Xu, G.L. and Zhu, W.M. (2007) Fabrication and tribological properties of carbon nanotubes reinforced Al composites prepared by pressureless infiltration technique. *Composites A*, **38**, 301–306.

28 Bakshi, S.R., Singh, V., Balani, K., McCartney, D.R., Seal, S. and Agarwal, A. (2008) Carbon nanotube reinforced aluminum composite coating via cold spraying. *Surface & Coatings Technology*, **202**, 5162–5169.

29 Zhong, R., Cong, H. and Hou, P. (2003) Fabrication of nano-Al based composites reinforced by single-walled carbon nanotubes. *Carbon*, **41**, 848–851.

30 Esawi, A.M. and El Borady, M.A. (2008) Carbon-nanotube-reinforced aluminum strips. *Composites Science and Technology*, **68**, 486–492.

31 Tjong, S.C. and Tam, K.F. (2006) Mechanical and thermal expansion of hipped Al-TiB$_2$ composites. *Materials Chemistry and Physics*, **97**, 91–97.

32 Bahnu Prasad, V.V., Bhat, B.V., Mahajan, Y.R. and Ramakrishnan, R. (2002) Structure-property correlation in discontinuously reinforced aluminum matrix composites as a function of relative particle size ratio. *Materials Science and Engineering A*, **337**, 179–186.

33 Slipenyuk, A., Kuprin, V., Milman, Y., Spowart, J.E. and Miracle, D.B. (2004) The effect of matrix to reinforcement particle size ratio (PSR) on the microstructure and mechanical properties of a P/M AlCuMn/SiC$_p$ MMC. *Materials Science and Engineering A*, **381**, 165–170.

34 George, R., Kashyap, K.T., Rahul, R. and Yamdagni, S. (2005) Strengthening in carbon nanotube/aluminum (CNT/Al) composites. *Scripta Materialia*, **53**, 1159–1163.

35 Deng, C.F., Wang, Z.Z., Zhang, X.X. and Li, A.B. (2007) Processing and properties of carbon nanotubes reinforced aluminum composites. *Materials Science and Engineering A*, **444**, 138–145.

36 Esawi, A. and Morsi, K. (2007) Dispersion of carbon nanotubes (CNTs) in aluminum powder. *Composites A*, **38**, 646–650.

37 Morsi, K. and Esawi, A. (2007) Effect of mechanical alloying time and carbon nanaotube (CNT) content on the evolution of aluminum (Al)-CNT composite powders. *Journal of Materials Science*, **42**, 4954–4959.

38 Perez-Bustamante, R., Estrada-Guel, I., Antunez-Flores, W., Miki-Yoshida, M., Ferreira, P.J. and Martinez-Sanchez, R. (2008) Novel Al-matrix nanocomposites reinforced with multi-walled carbon nanotubes. *Journal of Alloys and Compounds*, **450**, 323–326.

39 Lee, S.W., Choi, H.J., Kim, Y. and Bae, D.H. (2007) Deformation behavior of nanoparticles/fiber-reinforced nanocrystalline Al-matrix composites. *Materials Science and Engineering A*, **449–451**, 782–784.

40 Noguchi, T., Magario, A., Fukazawa, S., Shimizu, S., Beppu, J. and Seki, M. (2004) Carbon nanotube/aluminum composites with uniform dispersion. *Materials Transactions (JIM)*, **45**, 602–604.

41 Yuuki, J., Kwon, H., Kawasaki, A., Magario, A., Noguchi, T., Beppu, J. and Seki, M. (2007) Fabrication of carbon nanotube reinforced aluminum composites by powder extrusion process.

Materials Science Forum, **534–536** (Pt2), 889–892.

42 Fogagnolo, J.B., Ruiz-Navas, E.M., Robert, M.H. and Torralba, J.M. (2003) The effects of mechanical alloying on the compressibility of aluminum composite powder. *Materials Science and Engineering A*, **355**, 50–55.

43 Fogagnolo, J.B., Velasco, F., Robert, M.H. and Torralba, J.M. (2003) Effect of mechanical alloying on the morphology, microstructure and properties of aluminum matrix composite powders. *Materials Science and Engineering A*, **342**, 131–143.

44 Suryanarayana, C. (2001) Mechanical alloying and milling. *Progress in Materials Science*, **46**, 1–184.

45 Kamrani, S., Simchi, A., Riedel, R. and Seyed Reihani, S.M. (2007) Effect of reinforcement volume fraction on mechanical alloying of Al-SiC nanocomposite powders. *Powder Metallurgy*, **50**, 276–282.

46 He, C., Zhao, N., Shi, C., Du, X., Li, J., Li, H. and Cui, Q. (2007) An approach to obtaining homogeneously dispersed carbon nanotubes in Al powders for preparing reinforced Al-matrix composites. *Advanced Materials*, **19**, 1128–1132.

47 Valiev, R.Z., Zehetbauer, M.J., Estin, Y., Hoeppel, H.W., Ivanisenko, Y., Hahn, H., Wilde, G., Roven, H.J., Sauvage, X. and Langdon, T.G. (2007) The innovation potential of nanostructured materials. *Advanced Engineering Materials*, **9**, 527–533.

48 Lee, S., Berbon, P.B., Furukawa, M., Horita, Z., Nemoto, M., Tsenev, N.K., Valiev, R.Z. and Langdon, T.G. (1999) Developing superplastic properties in aluminum alloy through severe plastic deformation. *Materials Science and Engineering A*, **272**, 63–72.

49 Saravanan, M., Pillai, R.M., Ravi, K.R., Pai, B.C. and Brahmakumar, M. (2007) Development of ultrafine grain aluminum-graphite metal matrix composite by equal channel angular pressing. *Composites Science and Technology*, **67**, 1275–1279.

50 Sabirov, I., Kolednik, O., Valiev, R.Z. and Pippan, R. (2005) Equal channel angular pressing of metal matrix composites: Effect on particle distribution and fracture toughness. *Acta Materialia*, **53**, 4919–4930.

51 Stolyarov, V.V., Zhu, Y.T., Lowe, T.C., Islamgaliev, R.K. and Valiev, R.Z. (2000) Processing nanocrystalline Ti and its nanocomposites from micrometer-sized Ti powder using high pressure torsion. *Materials Science and Engineering A*, **282**, 78–85.

52 Tokunaga, T., Kaneko, K. and Horita, Z. (2008) Production of aluminum-matrix carbon nanotube composite using high pressure torsion. *Materials Science and Engineering A*, **490**, 300–304.

53 Tokunaga, T., Kaneko, K., Sato, K. and Horita, Z. (2008) Microstructure and mechanical properties of aluminum-fullerene composite fabricated by high pressure torsion. *Scripta Materialia*, **58**, 735–738.

54 Housh, S., Mikucki, B. and Stevenson, A. (1990) Selection and application of magnesium and magnesium alloys, in *ASM Handbook: Properties and Selection: Nonferrous Alloys and Special-Purpose Materials*, vol. **2**, ASM International, Materials Park, Ohio, USA, DOI:10.1361/asmhba0001074.

55 Xie, W., Liu, Y., Li, D.S., Zhang, J., Zhang, Z.W. and Bi, J. (2007) Influence of sintering routes to the mechanical properties of magnesium alloy and its composites produced by PM technique. *Journal of Alloys and Compounds*, **431**, 162–166.

56 Ferkel, H. and Mordike, B.L. (2001) Magnesium strengthened by SiC nanoparticles. *Materials Science and Engineering A*, **298**, 193–199.

57 Cao, G., Konishi, H. and Li, X. (2008) Mechanical properties and microstructure of SiC-reinforced Mg-

(2,4)Al-1Si nanocomposites fabricated by ultrasonic cavitation based solidification processing. *Materials Science and Engineering A*, **486**, 357–362.

58 Honma, T., Nagai, K., Katou, A., Arai, K., Suganuma, M. and Kamado, S. Synthesis of high-strength magnesium alloy composites reinforced with Si-coated carbon nanofibers. *Scripta Materialia*, DOI:10.1016/jscriptamat.2008.11.024.

59 Hassan, S.F. and Gupta, M. (2006) Effect of type of primary processing on the microstructure, CTE and mechanical properties of magnesium/alumina nanocomposites. *Composite Structures*, **72**, 19–26.

60 Gupta, M., Lai, M.O. and Soo, C.Y. (1996) Effect of type of processing on the microstructural features and mechanical properties of Al-Cu/SiC metal matrix composites. *Materials Science and Engineering A*, **210**, 114–122.

61 Ho, K.F., Gupta, M. and Srivatsan, T.S. (2004) The mechanical behavior of magnesium alloy AZ91 reinforced with fine copper particulates. *Materials Science and Engineering A*, **369**, 302–308.

62 Goh, C.S., Wei, J., Lee, L.C. and Gupta, M. (2006) Development of novel carbon nanotube reinforced magnesium nanocomposites using the powder metallurgy technique. *Nanotechnology*, **17**, 7–12.

63 Shimizu, Y., Miki, S., Soga, T., Itoh, I., Todoroki, H., Hosono, T., Sakaki, K., Hayashi, T., Kim, Y.A., Endo, M., Morimoto, S. and Koide, A. (2008) Multiwalled carbon nanotube-reinforced magnesium alloy composites. *Scripta Materialia*, **58**, 267–270.

64 Ma, Z.Y. (2008) Friction stir processing technology: A review. *Metallurgical and Materials Transactions A-Physical Metallurgy and Materials Science*, **39**, 642–658.

65 Mishra, R.S., Ma, Z.Y. and Charit, I. (2003) Friction stir processing: a novel technique for fabrication of surface composite. *Materials Science and Engineering A*, **341**, 307–310.

66 Morisada, Y., Fujii, H., Nagaoka, T. and Fukusumi, M. (2006) MWCNTs/AZ31 surface composites fabricated by friction stir processing. *Materials Science and Engineering A*, **419**, 344–348.

67 Larsen, J.M., Russ, S.M. and Jones, J.W. (1995) An evaluation of fiber-reinforced titanium matrix composites for advanced high-temperature aerospace applications. *Metallurgical and Materials Transactions A-Physical Metallurgy and Materials Science*, **26**, 3211–3223.

68 Loretto, M.H. and Konitzer, D.G. (1990) The effect of matrix reinforcement reaction on fracture in Ti-6Al-4V-base composites. *Metallurgical and Materials Transactions A-Physical Metallurgy and Materials Science*, **21**, 1579–1587.

69 Kim, Y.J., Chung, H. and Kang, S.J. (2002) Processing and mechanical properties of Ti-6Al-4V/TiC *in situ* composite fabricated by gas-solid reaction. *Materials Science and Engineering A*, **333**, 343–350.

70 Eylon, D.H. and Froes, F.H. (1990) Titanium powder metallurgy products, in *ASM Handboos: Properties and Selection: Nonferrous Alloys and Special-Purpose Materials*, vol. **2**, ASM International, Materials Park, Ohio, USA, DOI:10.1361/asmhba0001083.

71 Tjong, S.C. and Mai, Y.W. (2008) Processing-structure-property aspects of particulate- and whisker-reinforced titanium matrix composites. *Composites Science and Technology*, **68**, 583–601.

72 Kuzumaki, T., Ujie, O., Ichinose, H. and Ito, K. (2000) Mechanical characteristics and preparation of carbon nanotube fiber reinforced Ti composite. *Advanced Engineering Materials*, **2**, 416–418.

73 Tjong, S.C. and Lau, K.C. (2000) Abrasive wear behavior of TiB_2 particle-reinforced copper matrix composites. *Materials Science and Engineering A*, **282**, 183–186.

74 Tjong, S.C. and Ma, Z.Y. (2000) High temperature creep behavior of *in-situ* TiB_2 particulate reinforced copper-based

composite. *Materials Science and Engineering A*, **284**, 70–76.
75 Tjong, S.C. and Wang, G.S. (2006) Low-cycle fatigue behavior of *in-situ* TiB$_2$/Cu composite prepared by reactive hot pressing. *Journal of Materials Science*, **41**, 5263–5268.
76 Chang, S.Y. and Lin, S. (1996) J. fabrication of SiCw reinforced copper matrix composite by electrodeless copper plating. *Scripta Materialia*, **35**, 225–231.
77 Yih, P. and Chung, D.D. (1996) Silicon carbide whisker copper-matrix composites fabricated by hot pressing copper coated whiskers. *Journal of Materials Science*, **31**, 399–406.
78 Lee, Y.F., Lee, S.L., Chuang, C.L. and Lin, J.C. (1999) Effect of SiC$_p$ reinforcement by electrodeless copper plating on properties of Cu/SiC$_p$ composites. *Powder Metallurgy*, **42**, 147–152.
79 Zhang, R., Gao, L. and Guo, J. (2004) Preparation and characterization of coated nanoscle Cu/SiC$_p$ composite particles. *Ceramics International*, **30**, 401–404.
80 Jang, Y., Kim, S., Jung, Y. and Lee, S. (2005) Tensile behavior of carbon nanofiber-reinforced Cu composites using the liquid infiltration process. *Metallurgical and Materials Transactions A-Physical Metallurgy and Materials Science*, **36**, 217–223.
81 Kim, K.T., Cha, S.I., Hong, Seong H. and Hong, Soon H. (2006) Microstructures and tensile behavior of carbon naotube reinforced Cu matrix nanocomposites. *Materials Science and Engineering A*, **430**, 27–33.
82 Nygren, M. and Shen, Z. (2003) On the preparation of bio-, nano- and structural ceramics and composites by spark plasma sintering. *Solid State Sciences*, **5**, 125–131.
83 Licheri, R., Fadda, S., Orru, R., Cao, G. and Buscaglia, V. (2007) Self-propagating high-temperature synthesis of barium titanate and subsequent densification by spark plasma sintering (SPS). *Journal of the European Ceramic Society*, **27**, 2245–2253.
84 Li, J.L., Bai, G.Z., Feng, J.W. and Jiang, W. (2005) Microstructure and mechanical properties of hot-pressed carbon nanotubes compacted by spark plasma sintering. *Carbon*, **43**, 2649–2653.
85 Sun, Y.P., Fu, K., Lin, Y. and Huang, W. (2002) Functionalized carbon nanotubes: Properties and applications. *Accounts of Chemical Research*, **35**, 1096–1104.
86 Cha, S.I., Kim, K.T., Arshad, S.N., Mo, C.B. and Hong, S.H. (2005) Extraordinary strengthening effect of carbon nanotubes in metal-matrix nanocomposites processed by molecular-level mixing. *Advanced Materials*, **17**, 1377–1381.
87 Kim, K.T., Cha, S.I. and Hong, S.H. (2007) Hardness and wear resistance of carbon nanotube reinforced Cu matrix nanocomposites. *Materials Science and Engineering A*, **449–451**, 46–50.
88 Kim, K.T., Eckert, J., Menzel, S.B., Gemming, T. and Hong, S.H. (2008) Grain refinement assisted strengthening of carbon nanotube reinforced copper matrix nanocomposites. *Applied Physics Letters*, **92**, 121901(1)–121901(3).
89 Kim, K.T., Cha, S.I., Gemming, T., Eckert, J. and Hong, S.H. (2008) The role of interfacial oxygen atoms in the enhanced mechanical properties of carbon nanotube-reinforced metal matrix nanocomposites. *Small*, **4**, 1936–1940.
90 Xu, L., Chen, X., Pan, W., Li, W., Yang, Z. and Pu, Y. (2007) Electrostatic-assembly carbon nanotube-implanted copper composite spheres. *Nanotechnology*, **18**, 435607(1)–435607(4).
91 Tjong, S.C. and Chen, H. (2004) Nanocrystalline materials and coatings. *Materials Science and Engineering R*, **45**, 1–88.
92 Dalla Torre, F., Van Swygenhoven, H. and Victoria, M. (2002) Nanocrystalline electrodeposited nickel: microstructure and tensile properties. *Acta Materialia*, **50**, 3957–3970.

93 Wei, H., Hibbard, G.D., Palumbo, G. and Erb, U. (2007) The effect of gauge volume on the tensile properties of nanocrystalline electrodeposits. *Scripta Materialia*, **57**, 996–999.

94 Qu, N.S., Zhu, D. and Chan, K.C. (2006) Fabrication of Ni-CeO$_2$ nanocomposite by electrodeposition. *Scripta Materialia*, **54**, 1421–1425.

95 Barnejee, S. and Chakravorty, D. (1998) Electrical resistivity of copper-silica nanocomposites synthesized by electrodeposition. *Journal of Applied Physics*, **84**, 799–805.

96 Barnejee, S., Ghosh, A.K. and Chakravorty, D. (1999) Electrical properties of iron-silica nanocomposites synthesized by electrodeposition. *Journal of Applied Physics*, **86**, 6835–6840.

97 Chai, G., Sun, Y., Sun, J.J. and Chen, Q. (2008) Mechanical properties of carbon nanotube-copper nanocomposites. *Journal of Micromechanics and Microengineering*, **18**, 035013–035018.

98 Lim, B., Kim, C.J., Kim, B., Shim, U., Oh, S., Sung, B.H., Choi, J.H. and Baik, S.H. (2008) The effects of interfacial bonding on mechanical properties of single-walled carbon nanotube reinforced copper matrix nanocomposites. *Nanotechnology*, **17**, 5759–5764.

99 Hong, Soon H., Cha, S.I., Kim, K.T. and Hong, Seong H. (2007) Method of producing metal nanocomposite powder reinforced with carbon nanotubes and the powder prepared thereby. U.S. Patent 7217311.

100 Jeon, Y.S., Byun, J.Y. and Oh, T.S. (2008) Electrodeposition and mechanical properties of Ni-carbon nanotube nanocomposite coatings. *Journal of Physics and Chemistry of Solids*, **69**, 1391–1394.

101 Choi, E.K., Lee, K.Y. and Oh, T.S. (2008) Fabrication of multi-walled carbon nanotubes-reinforced Sn nanocomposites for lead-free solder by an electrodeposition process. *Journal of Physics and Chemistry of Solids*, **69**, 1403–1406.

102 Dai, P.Q., Xu, W.C. and Huang, Q.Y. (2008) Mechanical properties and microstructure of nanocrystalline nickel-carbon nanotube composites produced by electrodeposition. *Materials Science and Engineering A*, **483–484**, 172–174.

103 Thiemig, D., Lange, R. and Bund, A. (2007) Influence of pulse plating parameters on the electrocodeposition of matrix metal nanocomposites. *Electrochimica Acta*, **52**, 7362–7371.

104 Hou, K.H., Hwu, W.H., Ke, S.T. and Ger, M.D. (2006) Ni-P-SiC composite produced by pulse and direct current plating. *Materials Chemistry and Physics*, **100**, 54–59.

105 Sun, Y., Sun, J.R. and Chen, Q.F. (2007) Mechanical strength of carbon nanotube-nickel nanocomposites. *Nanotechnology*, **18**, 505704(1)–505704(6).

106 Guo, C., Zuo, Y., Zhao, X., Zhao, J.M. and Xiong, J. (2007) The effects of pulse-reverse parameters on the properties of Ni-carbon nanotube composite coatings. *Surface & Coatings Technology*, **201**, 9491–9496.

107 Chen, X., Xia, J., Peng, J., Li, W. and Xie, S. (2000) Carbon-nanotube metal-matrix composites prepared by electrodeless plating. *Composites Science and Technology*, **60**, 301–306.

108 Wang, F., Arai, S. and Endo, M. (2005) The preparation of multi-walled carbon nanotubes with a Ni-P coating by an electrodeless deposition process. *Carbon*, **43**, 1716–1721.

109 Liu, H., Cheng, G., Zheng, R., Zhao, H. and Liang, C. (2006) Influence of synthesis process on preparation and properties of Ni/CNT catalyst. *Diamond and Related Materials*, **15**, 15–21.

110 Jeong, Y.J., Cha, S.I., Kim, K.T., Lee, K.H., Mo, C.B. and Hong, S.H. (2007) Synergistic strengthening effect of ultrafine-grained metals reinforced with carbon nanotubes. *Small*, **3**, 840–844.

111 Cha, S.I., Kim, K.T., Arshad, S.N., Mo, C.B., Lee, K.H. and Hong, S.H. (2006) Field-emission behavior of a carbon nanotube-implanted Co nanocomposite fabricated from pearl-necklace-structured carbon nanotube/Co powders. *Advanced Materials*, **18**, 553–558.

112 Goyal, A., Wiegand, D.A., Owens, F.J. and Iqbal, Z. (2006) Enhanced yield strength in iron nanocomposite with *in situ* grown single-wall carbon nanotubes. *Journal of Materials Research*, **21**, 522–528.

113 Goyal, A., Wiegand, D.A., Owens, F.J. and Iqbal, Z. (2007) Synthesis of carbide-free, high strength iron-carbon nanotube composite by *in situ* nanotube growth. *Chemical Physics Letters*, **442**, 365–371.

3
Physical Properties of Carbon Nanotube–Metal Nanocomposites

3.1
Background

In recent decades there has been a major advance in the development of metal-matrix microcomposites reinforced with carbonaceous fillers. Carbon fibers have many excellent thermal and mechanical properties which make them attractive components of strong, light-weight composites. Conventional PAN-based fibers with high tensile strength and low modulus find applications as reinforcement materials for structural composites. Pitch-based carbon fibers with lower tensile strength, high modulus, excellent thermal and electrical conductivity are ideal reinforcements for composite applications in which heat dissipation is crucial [1, 2]. Such materials can be used to design a thermal doubler for satellite radiator panels [3]. However, carbon fibers are chemically reactive with metals during the composite fabrication, particularly using the liquid metal process. Chemical reactions between carbon fiber and metal during composite processing would degrade the matrix/interface properties. Therefore, efforts have been made to improve the performances of metal-matrix composites (MMCs) by proper control of the interfacial characteristics [4].

Recently, thermal management within the overall design of electronic products is increasingly important with the increase of package density in semiconductor devices. The increasing heat flux densities from dense packaging and ineffective dissipation of the thermal energy can lead to premature failure of electronic devices. Therefore, thermal management in electronic devices becomes increasingly important for reliable and long life performances [5]. Heat dissipation is usually achieved by the use of heat sinks, heat spreaders and packaging materials. Many composite materials have been developed to transport heat within electronic devices and dissipate it into the ambient environment more efficiently [6]. These include MMCs reinforced with pitch-based carbon fibers and SiC particles. Table 3.1 lists typical physical properties of some commercially available Al-based composites used in electronic devices. For the purposes of comparison, the properties of pure Al, Cu and Si are also listed [7, 8]. In Table 3.1, AlSiC composite materials are commercial products of CPS Technologies Corporation in which aluminum alloy (A356) is reinforced with different volume contents of SiC particles. They are available at low cost with near net shape fabrication versatility. Their thermal

Carbon Nanotube Reinforced Composites: Metal and Ceramic Matrices. Sie Chin Tjong
Copyright © 2009 WILEY-VCH Verlag GmbH & Co. KGaA, Weinheim
ISBN: 978-3-527-40892-4

Table 3.1 Physical properties of inorganic materials used in electronic devices.

Material	Thermal conductivity (W m^{-1} K^{-1}, 25 °C)	CTE (10^{-6} °C^{-1})	Filler content (vol%)	Electrical resistivity (μΩ cm)
Pure Al Ref [7]	230	23	—	—
Pure Cu Ref [7]	400		—	—
Si Ref [7]	150	4.2		
AlSiC-9 Ref [8]	200	8	37	20.7
AlSiC-10 Ref [8]	200	9.77	45	20.7
AlSiC-12 Ref [8]	180	10.9	63	20.7
MetGraf 4-230 Ref [9]	in Plane (xy): 220-230 Thickness (z): 120	xy: 4 z: 24	—	—
Cu/CF Ref [7]	xy: 250 z: 170	xy:12 z:17	30	4
Cu/CF Ref [9]	xy: 210 z: 150	xy: 9 z:17	40	5

conductivity is in the range of 180–200 W m^{-1}K^{-1}, depending on the filler content. The thermal conductivity of Al-based matrix composites can be enhanced greatly by using graphitized carbon fiber. MetGraf 4–230 is the composite product of Metal Matrix Cast Composites Inc. having aluminum alloy (A356) matrix [9]. At present, many microcomposite materials are still under development to achieve even higher thermal conductivity and light weight. For instance, Prieto et al. fabricated Al–12Si and Ag–3Si based composites and their hybrids filled with different kinds of carbon structures: graphite flakes (GF), CF and diamond, using gas-pressure-assisted liquid metal infiltration [10]; the results are summarized in Table 3.2. Graphite

Table 3.2 Thermal properties of the composites obtained with the different reinforcements for Al–Si and Ag–Si alloys.

Alloy	Reinforcement	Volume fraction	CTE (10^{-6} °C^{-1})	Thermal conductivity (W m^{-1}K^{-1})
Al–12Si	90%GF + 10%CF	0.88	z: 24 xy: 3	xy: 367
Ag–3Si	90%GF + 10%CF	0.88	z: 21 xy: 3	xy: 548
Al–12Si	60%GF + 40%SiC	0.88	z: 11 xy: 7	xy: 368
Ag–3Si	63%GF + 37%SiC	0.88	z: 11 xy: 8	xy: 360
Al–12Si	CF	0.55	xyz: 2.8	xyz: 131
Al–12Si	Diamond	0.61	xyz: 11	xyz: 350

Reproduced with permission from [10]. Copyright © (2008) Elsevier.
Note: xy refers to the graphene planes, while the direction perpendicular ti it is denoted by z. xyz indicates that the property is isotropic.

flakes with platelets of 400 μm are recommended due to their low cost and excellent thermal conductivity.

The size of integrated circuits (IC) continues to decrease as the demand for IC in wireless communication and personal computers increases. Miniaturization of IC chips and increasing processing speed reduces the heat transfer surface area and increases power. Local overheating in the central processing unit of computers can cause the breakdown of the whole system. Accordingly, heat dissipation is the most common problem that limits the performance, reliability and miniaturization of microelectronics. Further, rapid technological advancement in nanotechnology has led to the development of novel nanoelectronic devices with multifunctional properties [11–13]. Conventional metal-matrix composites reinforced with large volume content of ceramic microparticulates cannot cope with the fast development of electronic devices. Thus, composites reinforced with a low loading level of carbon nanotubes (CNTs) show great promise for thermal management applications. At present, only polymer-based composites reinforced with CNTs have been developed into thermal interface materials (TIM) for heat dissipation in electronic components. For instance, Huang *et al.* fabricated a prototype TIM film consisting of aligned CNT arrays embedded in an elastomer [14]. They achieved a 120% enhancement of thermal conductivity in the nanocomposite film reinforced with 0.4 vol.% CNT. However, the thermal conductivity of such nanocomposite film ($1.21\,\mathrm{W\,m^{-1}K^{-1}}$) is far less small than the theoretical thermal conductivity of aligned nanotubes.

Various unique properties of CNTs have generated interest among many researchers to use them as nanofillers for metals. The Young's modulus of CNTs predicted from theoretical simulations and measured experimentally is of the order of 1000 GPa. The tensile strength of MWNTs determined experimentally is 150 GPa. Further, theoretical prediction reveals an extremely high value of $6000\,\mathrm{W\,m^{-1}\,K^{-1}}$ for the room temperature thermal conductivity of an isolated SWNT. The measured room temperature thermal conductivity of an individual MWNT is $3000\,\mathrm{W\,m^{-1}\,K^{-1}}$. The combination of high strength, stiffness, aspect ratio and flexibility make CNTs ideal reinforcement materials for high performance composites employed for structural engineering and space applications. Despite these advantages, research studies on the thermal and electrical conducting behaviors of metal/CNT nanocomposites are very scarce. Nevertheless, there are tremendous opportunities for materials scientists to develop novel metal-matrix nanocomposites with fascinating properties.

3.1.1
Thermal Response of Metal-Matrix Microcomposites

The thermal expansion behavior of ceramic particulate-reinforced MMCs is controlled by several factors including size, distribution and content of reinforcing particles. In general, internal stresses are created within the composite during the cooling from processing temperature due to the mismatch in thermal expansion of the constituents. The thermal expansion behavior of a composite is quantified by the coefficient of thermal expansion (CTE) defined as the change in unit length of a material due to a rise or drop in temperature. Mathematically, it can be written as:

$$\alpha = \frac{\Delta l}{l \Delta T} \tag{3.1}$$

where α is the CTE, Δl the thermal expansion displacement, l the original length and ΔT the temperature change.

The CTE of the microcomposites can be predicted from several thermo-elastic models such as Kerner, Turner and Schapery provided that the corresponding effective elastic moduli of the composite are known. The Kerner model assumes the reinforcement is spherical and wetted by a uniform layer of matrix. The composite is considered as macroscopically isotropic and homogeneous [15]. The CTE of the composite can be expressed as:

$$\alpha_c = \alpha_p + \alpha_m V_m + (\alpha_p - \alpha_m) V_p V_m \Lambda \tag{3.2}$$

$$\Lambda = \frac{K_p - K_m}{V_p K_p + V_m K_m + \frac{3}{4} K_p K_m G_m} \tag{3.3}$$

where V is the volume fraction, α the CTE of the component. The subscripts c, p and m represent the composite, particulate reinforcement and matrix, respectively. K and G are the bulk and shear modulus of the components of the composite, respectively and relate to the Young's modulus and the Poisson's ratio υ of isotropic materials by:

$$K = \frac{E}{3(1 - 2\upsilon)} \tag{3.4}$$

$$G = \frac{E}{2(1 + \upsilon)}. \tag{3.5}$$

The Turner model assumes homogeneous strain throughout the composite and that only uniform hydrostatic stresses exist in the phases [16]. The CTE of the composite is given by:

$$\alpha_c = \frac{\alpha_m V_m K_m + \alpha_p V_p K_p}{V_m K_m + V_p K_p}. \tag{3.6}$$

Shapery's model gives upper (+) and lower (−) limits on the CTE [17]. The upper limit of CTE can be expressed as:

$$\alpha_C^{(+)} = \alpha_p + (\alpha_m - \alpha_p) \frac{K_m (K_p - K_C^{(-)})}{K_C^{(-)} (K_p - K_m)} \tag{3.7}$$

where $K_C^{(-)}$ is Hashin and Shtrikman lower limit to the bulk modulus of the composite given by [18, 19]:

$$K_C^{(-)} = K_m + \frac{V_p}{\frac{1}{K_p - K_m} + \frac{V_m}{K_m + \frac{4}{3} G_m}}. \tag{3.8}$$

Thus, the lower limit on $K_C^{(-)}$ yields the upper limit on the composite CTE (and vice versa). Shapery demonstrated that the upper limit coincides with the expression derived by Kerner. The lower limit of Schapery and the Turner model describe composites with reinforcement forming a percolative interpenetrating network. It is noted that the theoretical CTE values predicted by these models do not closely match experimental results in some cases. This is because internal stresses are created within the composite as a result of the difference in CTEs between composite components. This leads to plastic yielding of the metal matrix of the composite.

More recently, Geffroy *et al.* determined the thermal conductivity of copper film reinforced with 30 and 40 vol.% carbon fibers (pitch type) [7]. The thermal conductivity and CTE properties of Cu/CF composites are strongly anisotropic as shown in Table 3.1. The anisotropic properties arise from the preferential orientation of carbon fibers in the composites. Moreover, the measured values of thermal conductivity parallel to the surface films agree reasonably with those predicted from the Hashin and Shtrikman model. However, the experimental thermal conductivity values perpendicular to surface films deviate markedly from theoretical predictions.

3.2
Thermal Behavior of Metal-CNT Nanocomposites

3.2.1
Aluminum-Based Nanocomposites

Aluminum is an ideal material for heat dissipation in microelectronic devices due to its reasonably good thermal conductivity; its shortcoming is its relatively high CTE value (Table 3.1). The incorporation of CNTs with low CTE into Al can effectively reduce its thermal expansion. Tang *et al.* investigated the thermal behavior of nanocrystalline aluminum reinforced with SWNTs of different volume fractions [20]. The nanocomposites were prepared by mixing Al nanopowders and purified SWNTs in alcohol under sonication. The mixture was dried, cold compacted into disks, followed by hot consolidation at 380 °C. Figure 3.1 shows the relative thermal expansion vs temperature plots for coarse-grained Al, single crystal silicon and Al/15 vol.%SWNT nanocomposite specimens. The dimensional changes of these specimens increase with increasing temperature. The difference between the composite and Si is about one fifth of that between the coarse-grained Al and Si. Thus, CNT addition improves the thermal stability of the nanocomposite considerably. The CTE vs temperature plots for coarse-grained Al, Si and Al/SWNT nanocomposites containing different filler contents are given in Figure 3.2. Apparently the CTEs of the nanocomposites decrease with increasing SWNT content. The CTE of the Al/15 vol% SWNT nanocomposite is about one third of that of nano-Al at the temperature range of 50–250 °C, indicating that the CNTs effectively restrict the thermal expansion of the matrix. Since SWNTs are very effective in reducing the CTE of aluminum, the resulting nanocomposite shows great promise for electronic packaging applications. Very recently, Deng *et al.* also reported a beneficial effect of MWNTs in reducing the

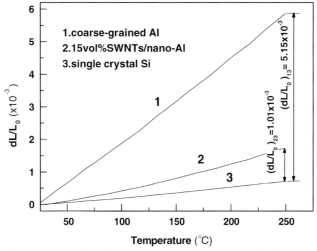

Figure 3.1 Linear thermal expansion curves of coarse-grained Al, single crystal silicon and Al/15 vol% SWNT nanocomposite specimens. Reproduced with permission from [20]. Copyright © (2004) Elsevier.

CTE of 2024 Al alloy. The addition of 1 wt% MWNT to the 2024 Al decreases its CTE by nearly 11% at 50 °C [21].

Figure 3.3 shows the experimental and theoretical plots of CTE vs CNT content for the Al/SWNT nanocomposites. The measured CTE values of the nanocomposites are

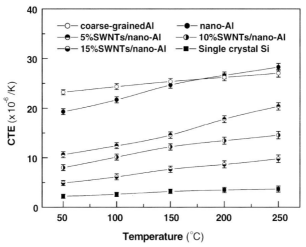

Figure 3.2 Temperature dependence of CTE of coarse-grained Al, single crystal silicon coarse-grained Al, single crystal silicon and Al/15 vol% SWNT nanocomposite specimens and Al/15 vol% SWNT nanocomposite specimens. Reproduced with permission from [20]. Copyright © (2004) Elsevier.

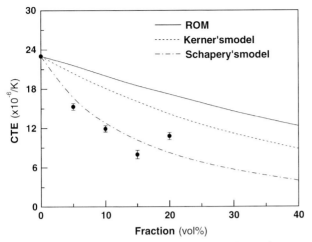

Figure 3.3 Comparison between experimental CTE values and theoretical predictions for Al/SWNT nanocomposites. Reproduced with permission from [20]. Copyright © (2004) Elsevier.

considerably lower than those predicted from the rule of mixtures (ROM) and Kerner models, but are closer to those predicted by the Schapery equation. As mentioned above, the lower limit of the Schapery model determines the CTE of composites having an interconnected filler network. For the Al/SWNT nanocomposites, the large aspect ratio of conducting SWNTs facilitates formation of interconnected filler network. The deviation of the measured CTE values for the nanocomposites with fillers ≥15 vol% from Schapery model is due to the agglomeration of SWNTs.

Jang et al. determined the thermal conductivity of 2024 Al alloy reinforced with 3 wt% carbon nanofibers of 120 nm diameter [22]. The nanocomposites were prepared by ball milling followed by hot isostatic pressing. Figure 3.4 shows thermal conductivity vs temperature plots for the 2024 Al alloy and 2024 Al/3 wt% CNF nanocomposite. The thermal conductivity of the 2024 Al/3 wt%CNF nanocomposite increases linearly with temperature between 15 and 300 K. The incorporation of carbon nanofibers with very high thermal conductivity (1260–2000 W m^{-1}K^{-1}) into 2024 Al alloy improves its thermal conductivity by ∼70% at room temperature. The thermal conductivity of 2024 Al/3 wt%CNF nanocomposite at room temperature (∼260 W m^{-1}K^{-1}) is comparable to that of copper based composite reinforced with 30% carbon fiber (250 W m^{-1}K^{-1}) as listed in Table 3.1.

3.2.2
Tin-Based Nanosolder

Advances in IC technology have led to the development of miniaturized semiconductor devices with advanced functional properties. Successful development of these devices requires proper understanding and tackling of thermal management,

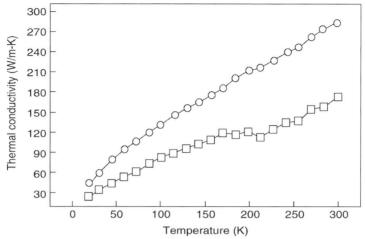

Figure 3.4 Thermal conductivity of the 2024 Al alloy and 2024 Al/3 wt% CNF nanocomposite between 15 and 300 K. ○: 2024 Al/3 wt% CNF; □: 2024 Al. Reproduced with permission from [22]. Copyright © (2007) Sage Publications.

materials selection and electronic packaging issues. Tin–lead solder alloy (Sn–37%Pb) is often used to connect chips to the packaging substrates in flip chip technology and surface mount technology due to its excellent solderability and low melting point (183 °C). However, the microstructure of the Sn–37%Pb alloy coarsens at elevated temperature, leading to the degradation of mechanical property and reliability of the joint. Further, the Sn–37%Pb alloy possesses environmental hazard because of its high lead content. Therefore, the electronic industry is seriously searching for lead-free alloy (e.g. SnAgCu solder) used for interconnects. As solder joints become increasingly miniaturized to meet the challenges of electronic packaging, the reliability of the flip chip solder joints has attracted considerable attention [23]. To improve the reliability efficiency of solder joints, second phase particles and reinforcing fillers are added to the solder alloys [24, 25]. The second phase particles or fillers can enhance the strength, suppress grain boundary sliding and grain growth of solder alloys. It is considered that CNT with superior mechanical, electrical and thermal properties is an excellent candidate reinforcement material for a nanocomposite solder [26–30].

Mohan Kumar et al. from the National University of Singapore studied the effects of single-wall CNT additions on the thermal and mechanical properties of Sn–37%Pb and lead-free Sn–3.8%Ag–0.7%Cu solder alloys [25–27]. They prepared nanocomposite solders by powder metallurgy followed by sintering, and reported that the SWNT additions improve the mechanical strength and stiffness of solder alloys. Moreover, the melting temperature of the solder alloy decreased slightly on incorporation of CNTs as revealed by differential scanning calorimetric (DSC) curves (Figure 3.5(a) and (b)). This implies that there is no need to adjust existing soldering practices when using CNT-reinforced solder alloy. The CTE of the Sn–37%Pb solder

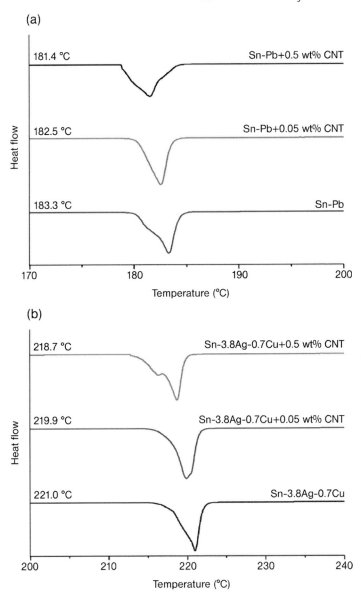

Figure 3.5 DSC curves showing the shift of melting point of (a) 63% Sn–37%Pb and (b) lead free Sn–3.8%Ag–0.7%Cu solder alloys to lower temperatures by adding single-wall CNTs. Reproduced with permission from [27]. Copyright © (2006) Elsevier.

can also be reduced to the value close to that of the organic substrate or printed circuit board (PCB) by adding 0.3–0.5% SWNT (Table 3.3). As recognized, the main reliability issue of the flip chip packaging technique arises from the thermal stresses induced by CTE mismatch between the silicon chip and PCB. The low CTE value of

Table 3.3 CTE values of the Sn–37%Pb alloy and its CNT-reinforced composites.

Material	CTE ($10^{-6}\,°C^{-1}$)
63%Sn–37%Pb	25.8 ± 1.2
63%Sn–37%Pb/0.01% SWNT	25.2 ± 0.9
63%Sn–37%Pb/0.03% SWNT	24.6 ± 1.1
63%Sn–37%Pb/0.05% SWNT	23.4 ± 0.7
63%Sn–37%Pb/0.08% SWNT	21.2 ± 1.3
63%Sn–37%Pb/0.1% SWNT	20.8 ± 1.4
63%Sn–37%Pb/0.3% SWNT	19.8 ± 1.1
63%Sn–37%Pb/0.5% SWNT	19.2 ± 0.9
PCB	18
SWNT	−1.6

Reproduced with permission from [27]. Copyright © (2006) Elsevier.

composite solder reinforced with nanotubes which closely matches that of the PCB can address the major reliability concern [29].

3.3
Electrical Behavior of Metal-CNT Nanocomposites

Up till now, the electrical behavior only of Al-CNT nanocomposites has been reported in the literature. Xu et al. measured the electrical resistivity of Al/MWNT nanocomposites filled with 1, 4 and 10 wt% CNTs from room temperature down to 4.2 K [31]. The composites were prepared by hand grinding MWNTs with Al powder, followed by hot pressing. Figure 3.6 shows the variation of electrical resistivity with temperatures for the Al/MWNT nanocomposites. The resistivity of all composites decreases linearly with temperature from 295 down to ∼80 K. At temperatures ≤80 K, the resistivity of the composites approaches zero. The loss of electrical resistance of nanocomposites at low temperatures is somewhat similar to that of superconducting materials. The mechanism for zero resistance of the nanocomposites at low temperatures could be attributed to the ballistic conduction of CNTs. MWNTs conduct current ballistically and do not dissipate heat. It has been theoretically predicted that ballistic conduction occurs without resistance in CNTs due to the disappearance of scattering. When scattering occurs during electric conduction very few ballistics are produced [32].

Xu et al. demonstrated that the electrical resistivity of Al ($3.4\,\mu\Omega\,cm$) increases to $4.9\,\mu\Omega\,cm$ by adding 1 wt% MWNT [31]. The resistivity further increases to $6.6\,\mu\Omega\,cm$ when 4 wt% MWNT is added. The reported experimental resistivity of individual CNT is in the order of 10^{-6}–$10^{-4}\,\Omega\,cm$, being much lower than that of aluminum. They attributed the increase in resistivity of the nanocomposites to agglomeration of CNTs at grain boundaries. The agglomerates enhance the scattering of the charge carrier at grain boundaries, thereby reducing the conductivity. Similarly, the resistivity of the 2024 Al/3 wt% CNF nanocomposite also shows a linear

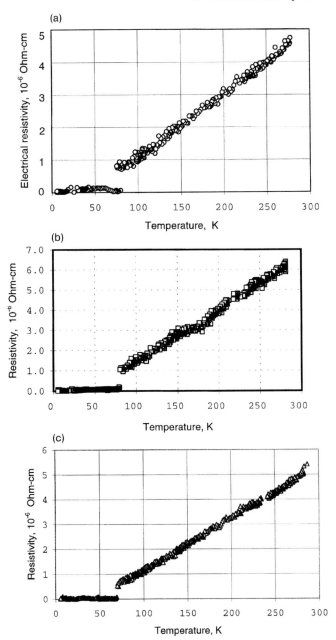

Figure 3.6 Dependence of electrical resistivity on temperature for (a) Al/1 wt% MWNT, (b) Al/4 wt% MWNT and (c) Al/10 wt% MWNT nanocomposites. Reproduced with permission from [31]. Copyright © (1999) Elsevier.

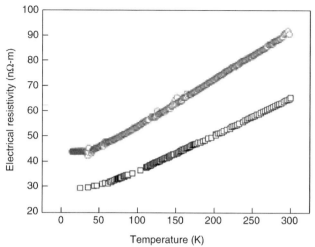

Figure 3.7 Electrical resistivity of the 2024 Al alloy and 2024 Al/3 wt% CNF nanocomposite between 15 and 300 K. ○: 2024 Al/3 wt% CNF; □: 2024 Al. Reproduced with permission from [22]. Copyright © (2007) Sage Publications.

decreasing trend with temperature as the temperature decreases (Figure 3.7). The resistivity of nanocomposite is higher than that of the 2024 Al alloy. Agglomeration of carbon nanofibers does not occur for this MA-prepared composite. Much more work is needed to elucidate this problem.

Nomenclature

α	Coefficient of thermal expansion
E	Young's modulus
G	Shear modulus
K	Bulk modulus
Δl	Thermal expansion displacement
l	Original length of specimen
ΔT	Temperature change.
υ	Poisson's ratio
V	Filler volume fraction

References

1 Callego, N.C. and Edie, D.D. (2001) Structure-property relationships for high thermal conductivity carbon fibers. *Composites A*, **32**, 1031–1038.

2 Glatz, J. and Vrable, D.L. (1993) Applications of advanced composites for satellite packaging for improved electronic component thermal

management. *Acta Astronautica*, **29**, 527–535.
3 Banisaukas, J.J., Shioleno, M.A., Levan, C.D., Rawal, S.P., Silverman, E.M. and Watts, R.J. (2005) Carbon fiber composites for spacecraft thermal management opportunities. *AIP Conference Proceedings*, **746**, 10–21.
4 Seong, H.G., Lopez, H.F., Robertson, D.P. and Rohatgi, P.K. (2008) Interface structure in carbon and graphite fiber reinforced 2014 aluminum alloy processed with active fiber cooling. *Materials Science and Engineering A*, **487**, 201–209.
5 Luedtke, A. (2004) Thermal management materials for high-performance applications. *Advanced Engineering Materials*, **6**, 142–144.
6 Zweben, C. (1998) Advances in composite materials for thermal management in electronic packaging. *Journal of Metals*, **50** (6), 47–51.
7 Geffroy, P.M., Chartier, T. and Silvain, J.F. (2007) Preparation by tape casting and hot pressing of copper carbon composite films. *Journal of the European Ceramic Society*, **27**, 291–299.
8 CPS Technologies Corporation, (2008) Materials Data Sheet, Norton, MA, USA. http://www.alsic.com.
9 Cornie, J.A., Zhang, S., Desberg, R. and Ryals, M. (2003) Discontinuous graphite reinforced aluminum and copper alloys for high thermal conductivity: Thermal expansion matched thermal management applications, http://www.mmccinc.com/graphitereinforcedalcu.pdf.
10 Prieto, R., Molina, J.M., Narciso, J. and Louis, E. (2008) Fabrication and properties of graphite flakes/metal composites for thermal management applications. *Scripta Materialia*, **59**, 11–14.
11 Tans, S.J., Verschueren, A.R. and Dekker, C. (1998) Room-temperature transistor based on a single carbon nanotube. *Nature*, **393**, 49–52.
12 Prinz, V.Y. (2003) A new concept in fabricating building blocks of nanodevices and nanomechanic devices. *Microelectronic Engineering*, **69**, 466–475.
13 Habenicht, B.F. and Prezhdo, O.V. (2008) Nanotubes devices: Watching electrons in real time. *Nature Nanotechnology*, **3**, 190–191.
14 Huang, H., Liu, C., Wu, Y. and Fan, S. (2005) Aligned carbon nanotube composite films for thermal management. *Advanced Materials*, **17**, 1652–1656.
15 Kerner, E. (1956) The elastic and thermoelastic properties of composite media. *Proceedings of the Physical Society B*, **69**, 808–813.
16 Turner, P. (1946) Thermal-expansion stresses in reinforced plastics. *Journal of Research, National Bureau of Standards*, **37**, 239–250.
17 Schapery, R. (1968) Thermal expansion coefficients of composite materials based on energy principles. *Journal of Composite Materials*, **2**, 380–404.
18 Hashin, Z. and Shtrickman, S. (1962) On some variational principles in anisotropic and nonhomogeneous elasticity. *Journal of the Mechanics and Physics of Solids*, **10**, 335–342.
19 Hashin, Z. and Shtrickman, S. (1963) A variational approach to the theory of the elastic behavior of multiphase elasticity. *Journal of the Mechanics and Physics of Solids*, **11**, 127–140.
20 Tang, Y., Cong, H., Zhong, R. and Cheng, H.M. (2004) Thermal expansion of a composite of single-walled carbon nanotubes and nanocrystalline aluminum. *Carbon*, **42**, 3251–3272.
21 Deng, C.F., Ma, Y.X., Zhang, P., Zhang, X.X. and Wang, D.Z. (2008) Thermal expansion behaviors of aluminum composite reinforced with carbon nanotubes. *Materials Letters*, **62**, 2302–2303.
22 Jang, J.H. and Han, K.S. (2007) Fabrication of graphite nanofibers reinforced metal matrix composites by powder metallurgy and their mechanical and physical characteristics. *Journal of Composite Materials*, **41**, 1431–1443.

23 Zeng, K. and Tu, K.N. (2002) Six cases of reliability study of Pb-free solder joint in electronic packaging technology. *Materials Science and Engineering R*, **38**, 55–105.

24 Lee, J.H., Park, D., Moon, J.T., Lee, Y.H., Shin, D.H. and Kim, Y.S. (2000) Reliability of composite solder bumps produced by *in-situ* process. *Journal of Electronic Materials*, **29**, 1264–1269.

25 Hwang, S.Y., Lee, J.W. and Lee, Z.H. (2002) Microstructure of a lead-free composite solder produced by an *in-situ* process. *Journal of Electronic Materials*, **31**, 1304–1308.

26 Nai, S.M., Wei, J. and Gupta, M. (2006) Lead-free solder reinforced with multiwalled carbon nanotubes. *Journal of Electronic Materials*, **35**, 1518–1522.

27 Mohan Kumar, K., Kripesh, V., Shen, L. and Tay, A.A. (2006) Study on the microstructure and mechanical properties of a novel SWCNT-reinforced solder alloy for ultra-fine pitch applications. *Thin Solid Films*, **504**, 371–378.

28 Mohan Kumar, K., Kripesh, V. and Tay, A.A. (2008) Single-wall carbon nanotube (SWCNT) functionalized Sn-Ag-Cu lead-free composite solders. *Journal of Alloys and Compounds*, **450**, 229–237.

29 Mohan Kumar, K., Kripesh, V. and Tay, A.A. (2008) Influence of single-wall carbon nanotube addition on the microstructural and tensile properties of Sn-Pb solder alloy. *Journal of Alloys and Compounds*, **455**, 148–158.

30 Choi, E.K., Lee, K.Y. and Oh, T.S. (2008) Fabrication of multiwalled carbon nanotubes-reinforced nanocomposites for lead-free solder by electrodeposition process. *Journal of Physics and Chemistry of Solids*, **69**, 1403–1406.

31 Xu, C.L., Wei, B.Q., Ma, R.Z., Liang, J., Ma, X.K. and Wu, D.H. (1999) Fabrication of aluminum-carbon nanotube composites and their electrical properties. *Carbon*, **37**, 855–858.

32 Chico, L., Benedict, L.X., Louie, S.G. and Cohen, M.L. (1996) Quantum conductance of carbon nanotubes with defects. *Physical Review B-Condensed Matter*, **54**, 2600–2606.

4
Mechanical Characteristics of Carbon Nanotube–Metal Nanocomposites

4.1
Strengthening Mechanism

The behavior of carbon nanotubes (CNTs) under tensile stress is expressed in terms of important parameters such as Young's modulus, tensile strength and elongation. The Young's modulus of CNTs predicted from theoretical simulations and measured experimentally is of the order of 1000 GPa. The tensile strength of MWNTs determined experimentally is 150 GPa. The combination of high strength, stiffness, aspect ratio and flexibility make CNTs ideal reinforcement materials for metals/alloys. Moreover, the excellent thermal and electrical conductivities of CNTs facilitate development of novel functional composites for advanced engineering applications. Recent studies have demonstrated significant improvements in the strength, Young's modulus and tensile ductility of metals/alloys reinforced with CNTs [Chap. 2, Ref. 46, Chap. 2, Ref. 62, Chap. 2, Ref. 81, 1]. It is considered that the strengthening mechanism is associated with load transfer from the metal matrix to the nanotubes. In general, micromechanical models are widely adopted by materials scientists to predict the tensile behavior of microcomposites reinforced with short fibers, whiskers and particulates. As CNTs exhibit large aspect ratios, we may ask whether the theories of composite mechanics and micromechanical models can be used to explain the mechanical properties of metal-CNT nanocomposites. Up till now, the principles of the mechanics of nanomaterials (nanomechanics) are in the early stages of development [2–5]. The development of CNT–metal nanocomposites still faces obstacles due to the lack of basic understanding of the origins of strengthening, stiffening and toughening and the matrix–nanotube interfacial issues.

The structure–property relationships of metal-matrix microcomposites are well recognized and reported. Several factors are known to contribute to the increments in yield strength and stiffness of metal-matrix microcomposites. These include effective load transfer from the matrix to the reinforcement, increasing dislocation density, homogenous dispersion of fillers and refined matrix grain size. Micromechanical models such as Cox shear-lag and Halpin–Tsai are often used to predict the stiffness and strength of discontinuously short-fiber-reinforced composites.

Experimental results of tensile measurements are then compared or correlated with such theoretical models.

The micromechanical models for discontinuously fiber-reinforced composites are briefly discussed here. The mechanical properties of such composites are mainly controlled by the stress transfer between the reinforcing fibers and the metal matrix at the interface. The strengthening mechanism through load transfer from the matrix to the reinforcement can be estimated from the shear-lag model. Cox originally proposed this approach where discontinuous fibers were embedded in an elastic matrix with a perfectly bonded interface and loaded in tension along the fiber direction [6]. Further, load transfer depends on the interfacial shear stress between the fiber and the matrix. The Cox model incorporates the aspect ratio ($S = l/d$ where l is the fiber length and d is the fiber diameter) of the reinforcement into the rule of mixtures. The Young's modulus of the composite then takes the following form:

$$E = \eta_l E_f V_f + E_m(1-V_f) \tag{4.1}$$

where E_f is the fiber modulus, E_m the matrix modulus, V_f the fiber volume fraction and η_l is defined by the following equation:

$$\eta_l = \left[1 - \frac{\tan h(\beta S)}{\beta S}\right] \tag{4.2}$$

β represents a group of dimensionless constants:

$$\beta = \sqrt{\frac{2E_m}{E_f(1+\upsilon_m)\ln(1/V_f)}} \tag{4.3}$$

where υ_m is the Poisson's ratio of matrix. The Cox model gives a poor prediction of the strengthening effect of discontinuously reinforced composites because the contribution of the load transfer effect at the fiber ends is not known. Nardone and Prewo [7, 8] modified the shear-lag model by taking the effect of tensile load transfer at the short fiber ends into consideration. Accordingly, the yield strength of a composite is expressed as:

$$\sigma_c = \frac{V_f \sigma_m S}{2} + \sigma_m \tag{4.4}$$

where V_m denotes the matrix volume fraction and σ_m the matrix yield strength

For the fiber with a misorientation angle θ with respect to the loading axis, Ryu et al. [9] demonstrated that the S parameter should be replaced by an effective aspect ratio (S_{eff}) for misaligned fiber. The modified yield strength of composite is given by the following equation:

$$\sigma_c = \frac{V_f \sigma_m S_{eff}}{2} + \sigma_m. \tag{4.5}$$

$$S_{eff} = S\cos^2\theta + \left(\frac{3\pi-4}{3\pi}\right)\left(1+\frac{1}{S}\right)\sin^2\theta. \tag{4.6}$$

Another approach for determining the elastic modulus of discontinuously aligned fiber-reinforced composite is to use the Halpin–Tsai equation [10]:

$$\frac{E_{CL}}{E_m} = \frac{1 + 2(l/d)\eta_L V_f}{1 - \eta_L V_f} \tag{4.7}$$

$$\frac{E_{CT}}{E_m} = \frac{1 + 2\eta_T V_f}{1 - \eta_T V_f} \tag{4.8}$$

where E_{CL} and E_{CT} are longitudinal and transverse composite modulus, respectively, η_L and η_T are given by the following equations:

$$\eta_L = \frac{(E_f/E_m) - 1}{(E_f/E_m) + 2(l/d)} \tag{4.9}$$

$$\eta_T = \frac{(E_f/E_m) - 1}{(E_f/E_m) + 2}. \tag{4.10}$$

For the randomly oriented short-fiber-reinforced composite, the elastic modulus of the composite can be calculated using the following equation [11]:

$$E_{random} = \frac{3}{8} E_{CL} + \frac{5}{8} E_{CT}. \tag{4.11}$$

The macroscale tensile test is one of the primary mechanical measurements commonly used to characterize the deformation of CNT–metal nanocomposites subjected to axial stress. Enhanced yield strength and elastic modulus have been observed in CNT–metal and CNT–ceramic systems during macroscale tensile deformation. The experimental data are compared with theoretical results predicted from micromechanical models of composites [12, 13]. For CNT–metal nanocomposites, effective load transfer from the matrix to the nanotubes is essential to achieve enhanced mechanical strength. From a microscopic viewpoint, strong interfacial bonding between the matrix and nanotube is needed to promote the load transfer. Fundamental understanding of the nature and mechanics of load transfer is crucial for making CNT–metal nanocomposites with tailored mechanical properties.

At present, direct tensile testing of CNT–metal nanocomposites in the nanoscale region is nonexistent in the literature. Only deformation of pristine CNTs at the nanoscale has been reported in the literature. Such nanoscale deformation was performed experimentally by means of atomic force microscopy (AFM) [Chap. 1, Ref. 144, 14, 15]. In AFM, individual nanotubes can be manipulated at the nanoscale using a cantilever with a sharp probe. When the tip comes close to the specimen surface, forces between the tip and the nanotube produce a deflection of the cantilever. The tip deflection can easily be detected by a photodetector that records the reflection of a laser beam focused on the top of the cantilever. Mechanical deformation can be evaluated from the force/tip-to-sample distance curves. Several approaches have been adopted to measure the nanoscale deformation of materials in AFM including nanoindentation, three-point bending and tension.

The resistance of materials to deformation (hardness) in very low volumes can be determined experimentally by nanoindentation. In this context, a small indentation is

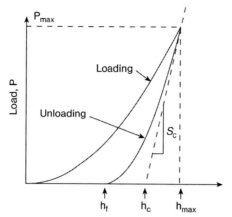

Figure 4.1 Load–displacement curve obtained from a nanoindentation test showing maximum indentation depth (h_{max}), contact depth (h_c) and final indentation depth (h_f).

made with a triangular diamond pyramid (Berkovich indenter). The measurement records the load and penetration depth continuously throughout the indentation loading and unloading cycles (Figure 4.1). The load–depth curves are then analyzed to yield hardness (H) and reduced elastic modulus (E_r). Oliver and Pharr [16, 17] analyzed the unloading portion of the load–depth curve and employed a fit to the unloading curve to obtain the contact stiffness (S_c) and contact depth (h_c) at the maximum applied load (P_{max}). Mathematically, S_c, H and E_r parameters can be expressed as:

$$S_c = \frac{2}{\sqrt{\pi}} E_r \sqrt{A} \tag{4.12}$$

$$\frac{1}{E_r} = \frac{(1-v_t^2)}{E_t} + \frac{(1-v_s^2)}{E_s} \tag{4.13}$$

$$H = \frac{P_{max}}{A} \tag{4.14}$$

where A is the contact area, which is a function of the contact depth, E_t and E_s are the elastic modulus, and v_t and v_s are the Poisson's ratio of the tip indenter and sample, respectively.

4.2
Tensile Deformation Behavior

4.2.1
Aluminum-Based Nanocomposites

To achieve an effective load transfer across the nanotube–matrix interface, the nanotubes must be distributed uniformly within the metal matrix. Therefore, prevention of agglomeration of CNTs is of particular importance for nanocomposites because most of their unique properties are only associated with individual nanotubes. As mentioned before, fabrication technologies play a dominant role in the uniform dispersion of nanotubes within the metal matrix. For CNT-Al nanocomposites prepared by conventional powder mixing, hot pressing and hot extrusion, the tensile strength of nanocomposites is inferior to that of pure Al due to the clustering of nanotubes [Chap. 2, Ref. 16]. This demonstrates the absence of the load transfer effect across the nanotube–matrix interface during tensile deformation.

George et al. [13] prepared the Al/MWNT and Al/SWNT nanocomposites by ball milling composite mixtures followed by sintering at 580 °C and hot extrusion at 560 °C. Pure aluminum was also fabricated under the same processing conditions. All these specimens were then subjected to tensile testing. The pure Al specimen exhibits low yield strength of 80 MPa and Young's modulus of 70 GPa. The experimental Young's modulus, yield strength and tensile strength of such nanocomposites are summarized in Table 4.1. Apparently, the elastic modulus of Al increases by 12 and 23% by adding 0.5 and 2 vol% MWNT, respectively. The elastic modulus values of the Al/0.5 vol% MWNT and Al/2 vol% MWNT nanocomposites correlate reasonably with those predicted from the shear-lag model assuming an aspect ratio of 100. Similarly, SWNT additions also enhance the stiffness and strength of aluminum as expected. The enhanced yield strength of nanotube-reinforced composites demonstrates that the applied load is effectively transferred across the nanotube–matrix interface.

Deng et al. prepared the 2024 Al/MWNT nanocomposites by PM followed by cold isostatic pressing and hot extrusion [Chap. 2, Ref. 35]. In the process, CNTs and 2024 Al powder (4.20 wt% Cu, 1.47 wt% Mg, 0.56 wt% Mn, 0.02 wt% Zr, 0.40 wt% Fe, 0.27 wt% Si) were dispersed in ethanol under sonication. The powder mixtures were dried, mechanically milled, followed by cold isostatic pressing and hot extrusion at 450 °C. Figure 4.2 shows the variations of relative density and Vickers hardness of 2024 Al/MWNT nanocomposites with nanotube content. The relative density and hardness increase with increasing nanotube content up to 1 wt%. At 2 wt% MWNT, the relative density and hardness decrease sharply due to the agglomeration of nanotubes. The Young's modulus and tensile strength of 2024 Al/MWNT nanocomposites also reach an apparent maximum at 1 wt% (Figure 4.3). The tensile

Table 4.1 Comparison of theoretical and experimental tensile data for Al/MWNT and Al/SWNT nanocomposites.

Materials	Shear lag Young's modulus (GPa)	Experimental Young's modulus (GPa)	Experimental yield strength (MPa)	Experimental ultimate tensile strength (MPa)
Al/0.5 vol%MWNT	74.31	78.1	86	134
Al/2 vol% MWNT	87.38	85.85	99	138
Al/1 vol% SWNT	79.17	70	79.8	141
Al/2 vol% SWNT	88.36	79.3	90.8	134

Reproduced with permission from [13]. Copyright © (2005) Elsevier.

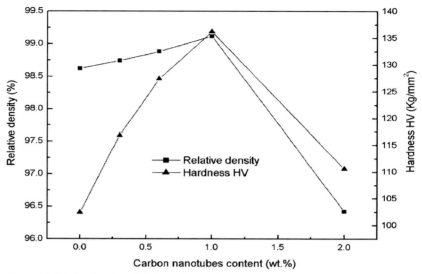

Figure 4.2 Relative density and Vickers microhardness as a function of carbon nanotube content for 2024 Al/MWNT nanocomposites. Reproduced with permission from [Chap. 2, Ref. 35]. Copyright © (2007) Elsevier.

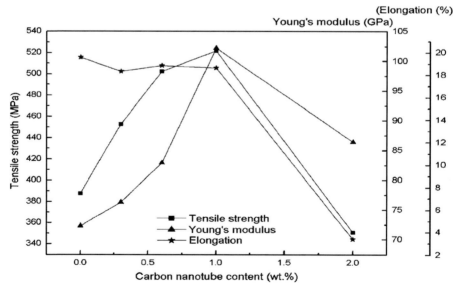

Figure 4.3 Tensile strength, Young's modulus and elongation vs carbon nanotube content for 2024 Al/MWNT nanocomposites. Reproduced with permission from [Chap. 2, Ref. 35]. Copyright © (2007) Elsevier.

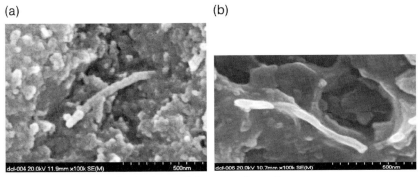

Figure 4.4 SEM micrographs showing tensile fracture surfaces of (a) 2024 Al/1 wt% MWNT and (b) 2024 Al/2 wt% MWNT nanocomposites. Reproduced with permission from [Chap. 2, Ref. 35]. Copyright © (2007) Elsevier.

elongation of nanocomposites remains nearly unchanged by adding nanotubes up to 1 wt%; this is due to the superior flexibility of CNTs. As a result, CNTs can bridge advancing crack during tensile deformation, thereby improving the tensile ductility of nanocomposites having nanotube content ≤1 wt% (Figure 4.4(a)). On the other hand, pull-out of nanotubes can be readily seen in the fracture surface of 2024 Al/2 wt % MWNT nanocomposite, indicating poor interfacial nanotube–matrix bonding strength (Figure 4.4(b)).

Nanocrystalline metals generally exhibit limited tensile plasticity due to a lack of work hardening capability. Therefore, cracks are readily initiated in nanocrystalline metals during the earlier stage of tensile deformation. Figure 4.5 shows typical tensile true stress–strain curve of nanocrystalline Al prepared by extruding ball-milled powders [18]. Apparently, nanocrystalline Al exhibits a high mechanical strength of

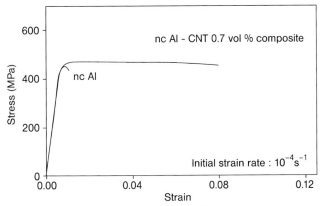

Figure 4.5 Tensile true stress–strain curves of nanocrystalline (nc) Al and its composite reinforced with 0.7 vol% SWNT. These materials were prepared by hot extrusion of ball-milled powders. Reproduced with permission from [18]. Copyright © (2007) Elsevier.

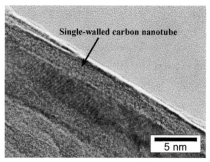

Figure 4.6 High-resolution TEM image of ncAl/0.7 vol% SWNT composite showing tightly bonded nanotubes–matrix interface. Reproduced with permission from [18]. Copyright © (2007) Elsevier.

460 MPa but extremely low tensile elongation (less than 1%). The tensile elongation of nanocrystalline Al can reach up to 8% by adding 0.7 vol% SWNT. The excellent flexibility of SWNTs resists the growth of necks in the nanocomposite during tensile deformation. The large aspect ratio and tightly bonded interface of nanotube with the matrix facilitate load transfer capacity in the composite (Figure 4.6).

4.2.2
Magnesium-Based Nanocomposites

Gupta and coworkers fabricated the Mg/MWNT nanocomposites containing nanotubes up to 0.30 wt% using conventional PM blending, sintering and hot extrusion [Chap. 2, Ref. 62]. The yield strength of magnesium increases from 127 to 133 MPa but the ultimate tensile strength slightly decreases from 205 to 203 MPa by adding 0.06 wt% MWNT. At 0.16 wt% MWNT, the yield strength increases to 138 MPa but the tensile strength remains almost unchanged (206 MPa). Further, agglomeration of MWNTs is observed in the fracture surface of Mg/0.3 wt% MWNT nanocomposite. As a result, little reinforcing effect is found by adding 0.3 wt% MWNT due to the nanotube clustering.

In order to disperse CNTs more uniformly into Mg-based composites during PM processing, it is necessary to shorten the length of CVD-grown nanotubes. Shimizu et al. used a high-speed blade cutting machine to reduce the length of CVD prepared MWNTs to an average length of ∼5 μm [Chap. 2, Ref. 63]. Damaged nanotubes were then introduced into the AZ91D matrix. The AZ91D/CNT nanocomposites were prepared by mechanical milling, hot pressing and hot extrusion. Table 4.2 lists the tensile properties of AZ91D and its nanocomposites. The elastic modulus, yield strength and tensile strength increase with increasing nanotube content up to 1 wt%. At 3–5 wt% MWNT the mechanical properties deteriorate due to the clustering of nanotubes.

Honma et al. also studied the tensile behavior of the AZ91D/CNF nanocomposites prepared by compocasting, squeeze casting and extrusion [Chap. 2, Ref. 58]. They reported that the yield stress and tensile strength of the nanocomposites

Table 4.2 Density and mechanical properties of AZ91D/MWNT nanocomposites.

Materials	Density (g cm^{-3})	Elastic modulus (GPa)	Tensile strength (MPa)	Yield stress (MPa)	Ductility (%)
AZ91D	1.80 ± 0.007	40 ± 2	315 ± 5	232 ± 6	14 ± 3
AZ91D/0.5%CNT	1.82 ± 0.008	43 ± 3	383 ± 7	281 ± 6	6 ± 2
AZ91D/1%CNT	1.83 ± 0.006	49 ± 3	388 ± 11	295 ± 5	5 ± 2
AZ91D/3%CNT	1.84 ± 0.005	51 ± 3	361 ± 9	284 ± 6	3 ± 2
AZ91D/5%CNT	1.86 ± 0.003	51 ± 4	307 ± 10	277 ± 4	1 ± 0.5

Reproduced with permission from [Chap. 2, Ref. 63]. Copyright © (2008) Elsevier.

increase slowly with increasing CNF content from 1.5 to 7.5 wt%. The yield stress and tensile strength of the AZ91D/1.5% CNF nanocomposite are 342 and 400 MPa, respectively. The yield stress and tensile strength of the AZ91D/7.5% CNF nanocomposite further increase to 416 and 470 MPa, respectively. The strengthening effect of nanofibers is pronounced and can be attributed to better dispersion of nanofibers in the magnesium alloy matrix. This arises from the employment of multiple casting processes and extrusion as well as the improvement in the wettability of nanofibers through the use of silicon coating. The mechanisms responsible for the strengthening of AZ91D/CNF nanocomposites include load transfer effect (Equation 4.4) and grain refinement strengthening (Hall–Petch relationship). Furthermore, the contribution of load transfer to the yield stress enhancement is twice of that of grain refinement.

In addition to squeeze casting, disintegrated melt deposition (DMD) seems to be an effective technique for depositing near net shape metal-matrix composites [Chap. 2, Ref. 61]. Gupta and workers used the DMD technique to fabricate the Mg/MWNT nanocomposites containing fillers from 0.3 up to 2 wt%. The resulting nanocomposites were then extruded [Chap. 2, Ref. 5, 19]. Figure 4.7 shows the stress–strain curves of pure Mg and Mg/MWNT nanocomposites. The density and mechanical properties of these specimens are listed in Table 4.3. Apparently, the density of nanocomposites remains almost unchanged by adding nanotubes up to 1.6 wt%. At 2 wt% MWNT, the density begins to decrease due to the formation of micropores. Tensile test data show that the yield and tensile strengths as well as tensile ductility of Mg-based nanocomposites increase with increasing nanotube content up to 1.3 wt%. At 1.6 and 2 wt% MWNT, the yield stress, tensile strength and ductility reduce considerably due to the agglomeration of nanotubes and micropore formation. The improvement in tensile ductility of nanocomposites with MWNT content ≤1.3 wt% is attributed to the high activity of the basal slip system and the initiation of prismatic $\langle a \rangle$ slip [Chap. 2, Ref. 5, 19]. Magnesium generally exhibits low tensile ductility due to its hexagonal close-packed (HCP) structure having only three independent slip system. MWNTs assist the activation of prismatic and cross-slip in the matrix during extrusion. Texture analysis reveals that the basal planes of magnesium matrix tend to align with the extrusion direction (Figure 4.8). Such dislocation slip behavior has been confirmed by transmission electron microscopyTEM observations.

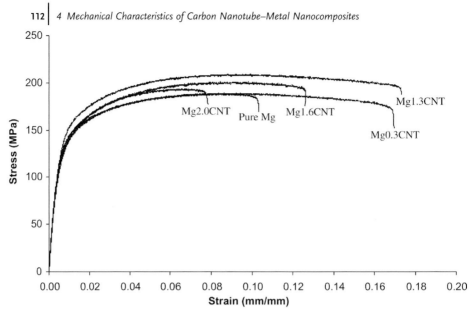

Figure 4.7 Tensile stress–strain curves of pure Mg and Mg/MWNT nanocomposites. Reproduced with permission from [19]. Copyright © (2008) Elsevier.

4.2.3
Copper-Based Nanocomposites

Copper nanopowders do not mixed properly with CNTs even after ball milling. For instance, Hong and coworkers fabricated Cu/5 vol% MWNT and Cu/10 vol% MWNT nanocomposites by spark plasma sintering of ball-milled Cu and nanotube powder mixtures and cold rolling. The microstructure of cold-rolled nanocomposites consists of a nanotube-rich surface region and a nanotube-free inner region [Chap. 2, Ref. 81]. Figure 4.9(a) shows the tensile stress–strain curves of pure Cu and its nanocomposites. Carbon nanotube additions improve the yield and tensile strengths of copper at the expense of tensile ductility. The variation of elastic modulus, yield and tensile strengths with nanotube content is shown in Figure 4.9(b). Careful examination of Figure 4.9(a) reveals the presence of two yield points owing to the

Table 4.3 Density and mechanical properties of Mg/MWNT nanocomposites.

Materials	Density (g cm^{-3})	0.2% Yield stress (MPa)	UTS (MPa)	Ductility (%)
Mg	1.738 ± 0.010	126 ± 7	192 ± 5	8.0 ± 1.6
Mg/0.3 wt% MWNT	1.731 ± 0.005	128 ± 6	194 ± 9	12.7 ± 2.0
Mg/1.3 wt% MWNT	1.730 ± 0.009	140 ± 2	210 ± 4	13.5 ± 2.7
Mg/1.6 wt% MWNT	1.731 ± 0.003	121 ± 5	200 ± 3	12.2 ± 1.72
Mg/2 wt% MWNT	1.728 ± 0.001	122 ± 7	198 ± 8	7.7 ± 1.0

Reproduced with permission from [Chap. 2, Ref. 5]. Copyright © (2006) Elsevier.

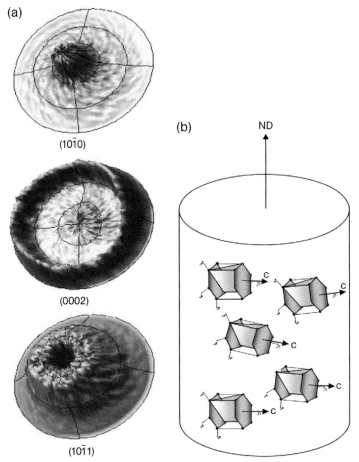

Figure 4.8 Texture analysis of Mg/1.3 wt% MWNT nanocomposite showing (a) $(10\bar{1}0)$, (0002) and $(10\bar{1}1)$ pole figures and (b) alignment of basal and non-basal planes according to the pole figure results. Reproduced with permission from [19]. Copyright © (2008) Elsevier.

inhomogeneous distribution of nanotubes. The primary yield strength can be estimated theoretically from modified shear-lag equations (Equations 4.5 and 4.6). The results are listed in Table 4.4. Apparently, modified shear-lag model predictions agree reasonably with the experimentally measured data.

To obtain homogeneous dispersion of nanotubes in a copper matrix, the molecular level mixing method was used to prepare the Cu/5 vol% MWNT and Cu/10 vol% MWNT nanocomposites. The resulting composite powders were then spark plasma sintered at 550 °C for 1 min [Chap. 2, Ref. 86, Chap. 2, Ref. 87, Chap. 2, Ref. 89]. Figure 4.10(a) shows compressive stress–strain curves of pure Cu and its nanocomposites. The variation of elastic modulus and yield strength with nanotube content is shown in Figure 4.10(b). The Cu/MWNT nanocomposites only exhibit one yield

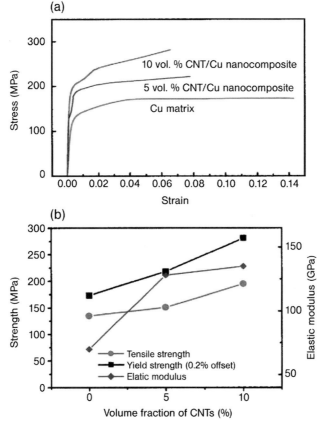

Figure 4.9 (a) Tensile stress–strain curves of pure Cu and Cu/MWNT nanocomposites. (b) Variation of elastic modulus, yield and tensile strengths with nanotube content. The nanocomposites were prepared by spark plasma sintering of ball-milled Cu nanopowder and MWNT mixture followed by cold rolling. Reproduced with permission from [Chap. 2, Ref. 81]. Copyright © (2006) Elsevier.

Table 4.4 Experimental and shear-lag predicted yield strength of Cu/MWNT nanocomposites.

Volume content of MWNT (%)	Volume content of Cu/MWNT composite region (%)	S_{eff} of Cu/MWNT composite region	Primary yield strength (MPa)	Theoretical primary yield strength (MPa)
5	6.3	3.2	149	151
10	12.7	4.8	197	180

Reproduced with permission from [Chap. 2, Ref. 81]. Copyright © (2006) Elsevier.

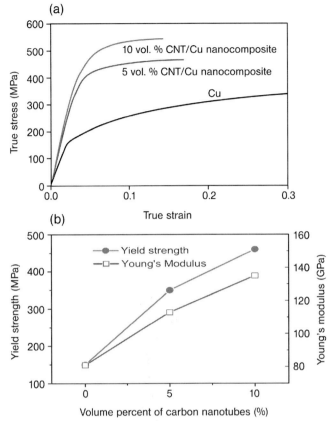

Figure 4.10 (a) Compressive true stress–strain curves of pure Cu, and Cu/MWNT nanocomposites prepared by molecular level mixing and spark plasma sintering. (b) Young's modulus and compressive yield strength vs carbon content for Cu/MWNT nanocomposites. Reproduced with permission from [Chap. 2, Ref. 86]. Copyright © (2005) Wiley-VCH Verlag GmbH.

point due to homogenous dispersion of nanotubes in the copper matrix. The yield strength of copper improves dramatically with the incorporation of nanotubes. The yield strength of Cu/5 vol% MWNT nanocomposite is 360 MPa, which is 2.4 times that of Cu (150 MPa). The compressive yield strength increases to 455 MPa when the nanotube content reaches 10 vol%, being 3 times that of Cu. This is attributed to the high load-transfer efficiency of nanotubes in the matrix as a result of strong interfacial bonding between the copper matrix and nanotubes through molecular level mixing.

From Equation 4.4, the strengthening efficiency of the reinforcement (R) can be expressed as $S/2$, thereby yielding:

$$\sigma_c = \sigma_m(1 + V_f R) \tag{4.15}$$

$$R = \frac{(\sigma_c - \sigma_m)}{V_f \sigma_m}. \tag{4.16}$$

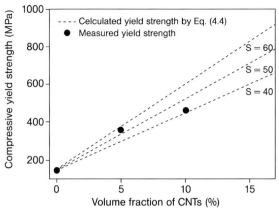

Figure 4.11 Experimental and theoretical yield strength of Cu/MWNT nanocomposites as a function of carbon nanotube content. Reproduced with permission from [Chap. 2, Ref. 89]. Copyright © (2008) Wiley-VCH Verlag GmbH.

The R values for the Cu/5 vol% MWNT and Cu/10 vol% MWNT nanocomposites are determined to be 14 and 20.3, respectively. These values are much higher than those of SiC particle and SiC whisker reinforcements. Figure 4.11 shows a comparison between theoretical and experimental compressive yield strength vs nanotube content for Cu/MWNT nanocomposites. Theoretical predictions are made assuming the aspect ratios of nanotubes to be 40, 50 and 60. Apparently, the measured yield strength agrees reasonably well with the estimated values from Equation 4.4 for the composites with aspect ratios of 40 and 50. In addition to the load transfer mechanism, metallurgical factors, such as refinement of matrix grains and higher dislocation density resulting from the coefficient of thermal expansion between the matrix and nanotubes, also contribute to the strength of nanocomposites [20].

4.2.4
Nickel-Based Nanocomposites

Electrodeposition is widely known as an effective technique for depositing dense nanocrystalline nickel coatings for mechanical characterization. Nanocrystalline nickel coatings generally exhibit much higher tensile strength and stiffness but poorer tensile ductility compared with microcrystalline Ni. It is considered that CNTs with superior tensile strength, stiffness and fracture strain can further improve the mechanical properties of Ni coatings. Table 4.5 lists the tensile properties of nanocrystalline Ni and nanocrystalline Ni/MWNT composite prepared by electrodeposition [Chap. 2, Ref. 100]. The hardness and tensile strength of Ni/MWNT composite are considerably higher than those of nanocrystalline Ni. The tensile elongation of nanocrystalline Ni decreases very slightly by incorporating MWNTs. It can be said that the tensile elongation is quite similar and not affected by the nanotube additions. Sun et al. [Chap. 2, Ref. 105] demonstrated that

Table 4.5 Mechanical properties of electrodeposited nanocrystalline Ni and Ni/MWNT composite.

Materials	Vickers Hardness	Tensile strength (MPa)	Elongation (%)
Nanocrystalline Ni	572	1162	2.39
Ni/MWNT Composite	645	1475	2.09

Reproduced with permission from [Chap. 2, Ref. 100]. Copyright © (2008) Elsevier.

the Ni/MWNT composite deposited by the pulse-reverse method exhibits higher tensile elongation (2.4%) than that of pure Ni (1.7%). The yield strength and tensile strength of Ni/MWNT composite are 1290 and 1715 MPa, respectively. By using single-walled nanotubes, the tensile strength of Ni/SWNT composite deposited under the same condition can reach as high as 2 GPa. The high reinforcing efficiency of CNTs is attributed to good interfacial bonding and stiffer nickel metal matrix.

4.3
Comparison with Nanoparticle-Reinforced Metals

Ceramic nanoparticle reinforcements can markedly increase the tensile strength and elastic modulus of metals more effectively than their microparticle counterparts. The improved mechanical properties of nanocomposites can be achieved by adding very low volume fractions of nanoparticles provided that the fillers are dispersed uniformly in the metal matrices of composites [Chap. 1, Ref. 2–Chap. 1, Ref. 4]. However, ceramic nanoparticles always agglomerate into large clusters during processing, particularly at higher filler content.

Figure 4.12 shows the tensile properties of PM Al reinforced with alumina nanoparticles (25 nm) of different volume fractions; and for comparison, the tensile behavior of Al reinforced with 10 vol% SiC (10 μm) particles are also shown. Apparently, the yield and tensile strengths of PM Al/Al$_2$O$_3$ nanocomposites increase with increasing alumina content up to 4 vol%, but thereafter level off due to severe clustering of nanoparticles. Figure 4.13(a) and (b) show typical TEM micrographs of the Al/1vol%Al$_2$O$_3$ and Al/4vol%Al$_2$O$_3$ nanocomposites. Clustering of alumina nanoparticles occurs when the filler content reaches 4 vol%. These clusters are mainly located at the grain boundaries of the aluminum matrix.

The aspect ratio of spherical nanoparticles is unity, thus the contribution from the load bearing transfer effect to the strength of composites is smaller compared with that of CNTs. In this case, Orowan stress plays a major role in strengthening composites reinforced with nanoparticles [21]. Mathematically, Orowan stress can be described by the following equation [22, 23]:

$$\tau = \frac{0.84\, MGb}{(L_m - d)} \tag{4.17}$$

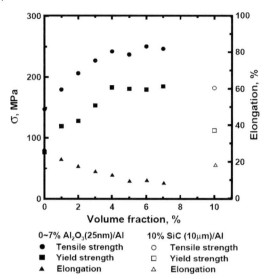

Figure 4.12 Tensile properties of Al/Al$_2$O$_3$ nanocomposites as a function of alumina volume fraction. The tensile properties of Al/10 vol% SiC (10 μm) microcomposite are also shown. Reproduced with permission from [Chap. 1, Ref. 4]. Copyright © (2004) Elsevier.

where τ is shear yield stress, G is shear modulus, M is Taylor factor (∼3), b is Burgers vector and L_m mean inter-particle distance given by:

$$L_m = \left(\frac{6V_f}{\pi}\right)^{-1/3} d. \tag{4.18}$$

Apparently, the yield stress of composites can be markedly enhanced by increasing the filler volume fraction and decreasing the particle diameter.

Recently, Gupta's group studied the mechanical properties of PM- and DMD-prepared Mg-based composites reinforced with different ceramic nanoparticles

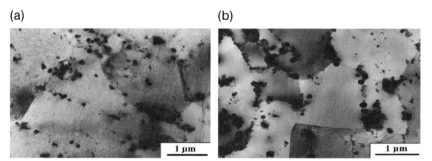

Figure 4.13 TEM micrographs of (a) Al/1vol%Al$_2$O$_3$ and (b) Al/4vol%Al$_2$O$_3$ nanocomposites. Reproduced with permission from [Chap. 1, Ref. 4]. Copyright © (2004) Elsevier.

Table 4.6 Tensile properties of PM and DMD prepared magnesium and its composites reinforced with ceramic nanoparticles.

Materials	E (GPa)	0.2% Yield stress (MPa)	UTS (MPa)	Elongation (%)
Pure Mg (PM) Ref [24]	41.2	132 ± 7	193 ± 2	4.2 ± 0.1
Mg/1.1 vol%Al$_2$O$_3$ (PM) Ref [24]	44.5	194 ± 5	250 ± 3	6.9 ± 1.0
Pure Mg (DMD) Ref [24]	42.8	97 ± 2	173 ± 1	7.4 ± 0.2
Mg/1.1 vol%Al$_2$O$_3$ (DMD) Ref [24]	52.7	175 ± 3	246 ± 3	14.0 ± 2.4
Mg/0.22 vol%Y$_2$O$_3$(DMD) Ref [25]	—	218 ± 2	277 ± 5	12.7 ± 1.3
Mg/0.66 vol%Y$_2$O$_3$(DMD) Ref [25]	—	312 ± 4	318 ± 2	6.9 ± 1.6
Mg/1.11 vol%Y$_2$O$_3$(DMD) Ref [25]	—	—	205 ± 3	1.7 ± 0.5

[24–26]. The tensile properties of Mg-based nanocomposites depend greatly on the processing techniques employed (Table 4.6). The yield and tensile strengths of Al/1.1 vol% Al$_2$O$_3$ (55 nm) nanocomposite prepared by the PM sintering technique is higher than of those fabricated by DMD process. This is attributed to the presence of porosities in the DMD-prepared nanocomposite. The strengthening mechanism for these nanocomposites derives mainly from the Orowan stress.

4.4 Wear

Wear is a loss of the surface material resulting from the relative motion between mating surfaces. The asperities of the two surfaces come into contact when two surfaces slide against each other under the application of an external load. The atoms of the softer material adhere to the asperities of the harder counterface. Contacting surface asperities cold weld together and form interatomic junctions across the interface. On further sliding, the junctions fracture and fragments detach from the adhering asperities, resulting in wear damage of softer surface. The wear damage of the specimen can be expressed in terms of the wear volume or weight loss. From Archard's theory for adhesive wear, the wear volume (V) of a material can be expressed as:

$$V = k\,PL/H \tag{4.19}$$

where k is the wear coefficient, P the applied load, L the sliding distance and H the hardness of wearing material [27]. Moreover, shearing of the junctions under applied force results in friction. With a multiasperity contacting mode, frictional force displays a linear relationship with normal force as given by the following equation:

$$F = \mu N \tag{4.20}$$

where F is the frictional force, N the normal force and μ the coefficient of friction.

As the asperities of the softer surface are deformed and fractured by the asperities of harder counterface, shear deformation is induced near the subsurface region of softer material. This gives rise to the generation and development of microcracks in this region. The degree of shear deformation depends on the applied load and sliding time. Fragments are detached from the softer specimen surface as wear debris. This type of material removal is termed as "delamination wear" [28]. In general, delamination wear is influenced to a large extent by the microstructure of tested specimens. The wear volume of composite materials depends not only on their hardness, but also on other intrinsic materials factors, such as dispersion state and distribution of fillers, size of fillers, interfacial bond strength, and extrinsic testing conditions. Inhomogeneous distribution of reinforcing particles can lead to extensive material loss due to disintegration of particle agglomerates.

The wear volume of materials at macro-levels is often determined from the pin-on-disk, ball-on-flat (plate) or block-on-ring tests. In a typical pin-on-disk measurement, the softer pin specimen is slid against a hard disk counterpart, resulting in weight loss of the pin. The measurement can also be carried out by sliding a hard pin against softer disk specimen. For the ball-on-flat test, a hard ball is rubbed against softer flat specimen, forming deep grooves in the specimen by a plowing action.

Carbon-based materials are widely known to be excellent solid lubricants. The metal/graphite composites are attractive materials for tribological applications in automotive engines due to the self-lubricating behavior of graphite [29–32]. In general, lubricants are used to minimize friction and reduce wear of the contact surfaces during sliding wear of component materials. The formation of the lubricating film prevents direct contact between the mating surfaces. However, it is necessary to apply lubricants periodically to the wear parts. Solid self-lubricating material like graphite can automatically induce formation of lubricating film during the wear process. The main problem of using graphite as a solid lubricant is the loss of composite strength during service application. This is because the brittle graphite fillers fracture readily under the application of an external load. In this regard, CNTs with exceptionally high strength and stiffness and superior flexibility are attractive fillers for composites used in tribological applications. The improvement in wear resistance of the metal-CNT composites is associated with the load-bearing capacity and low friction coefficient of CNTs [33].

The wear deformation and damage of metal-matrix composites are well established in the literature [34, 35]. Compared with wear damage at the microscale, the wear rate on the nanoscale is much smaller because the applied load is very low. Contact mode AFM is an effective tool for characterizing the nanowear behavior of materials at the atomic scale. An AFM with suitable hard tip mounted on a cantilever can stimulate or form a single asperity contact with the scanned surface. Since asperities exhibit all sizes and shapes, AFM tips with different radii ranging from about 20 to 70 nm have been produced. As the tip is pushed into the specimen surface, deflection of the cantilever occurs. A variety of forces (e.g. attractive, repulsive and adhesive) can be detected between the AFM tip and the specimen surface [36]. The wear behavior of materials can be determined by careful analysis of the force/tip-sample distance curves. Accordingly, AFM is widely used to measure nanoindentation, microscratching and nanowear behavior of solid surfaces and thin films [5, 36–38]. In general, adhesive wear mechanism still operates for metallic materials under light loads. Landman *et al.* used MD simulation and AFM to study the atomistic mechanism of contact formation between a hard nickel tip and soft gold sample [39]. Indentation of a gold surface by advancing a nickel tip resulted in the adhesion of gold atoms to the nickel tip and formation of a connective neck of atoms.

Interestingly, MD simulations show that CNTs can slide and roll against each other under the application of a shear force [40, 41]. The rolling of CNTs at the atomic level on the graphite surface has been verified experimentally by Falvo *et al.* using AFM [42]. They reported that rolling can occur only when both the nanotube and the underlying graphite have long-range order. Thus, a nanotube has preferred orientations on the graphite surface and rolling takes place when it is in atomic scale registry with the surface. This behavior is somewhat similar to the rolling of fullerenes that act like nanoscale ball bearings and solid lubricants [43, 44]. Despite the widespread use of AFM for characterizing the nanowear of metallic materials, little work has been done on nanowear of metal-CNT composites.

Up till now, there appears to be no reported work on the nanowear behavior of metal-CNT nanocomposites in the literature. Only wear properties of metal-CNT nanocomposites at macro-levels have been reported. Zhang and coworkers [Chap. 2, Ref. 28] investigated the dry friction and wear behaviors of Al-Mg/MWNT nanocomposites using a pin-on-disk test under a load of 30 N. The nanocomposites were prepared by liquid metallurgy route in which molten Al was infiltrated into CNT-Al-Mg preform under a nitrogen atmosphere. Figure 4.14 shows the variation of Brinell hardness with nanotube content for the Al-Mg/MWNT nanocomposites. The hardness of composites increases almost linearly with increasing nanotube content up to 10 vol% and then saturates with further increasing filler volume content. This leads to

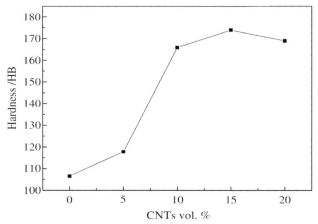

Figure 4.14 Brinell hardness (HB) vs carbon nanotube volume content for Al-Mg/MWNT nanocomposites. Reproduced with permission from [Chap. 2, Ref. 28]. Copyright © (2007) Elsevier.

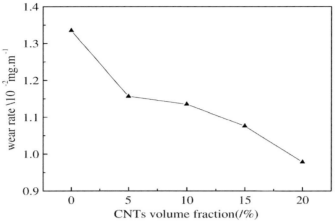

Figure 4.15 Wear rate vs carbon nanotube volume content for Al-Mg/MWNT nanocomposites at a sliding velocity of 0.1571 m/s under a load of 30 N. Reproduced with permission from [Chap. 2, Ref. 28]. Copyright © (2007) Elsevier.

a decrease in the wear rate of nanocomposites with increasing filler content (Figure 4.15). An increase in the volume content of CNTs minimizes the direct contact between the aluminum matrix and its sliding disk counterpart. This modifies the friction due to the self lubrication of nanotubes. As a result, the friction coefficient of nanocomposites can reach a value as small as 0.105 by adding 15 vol% MWNT (Figure 4.16). In another study, they prepared Cu/MWNT nanocomposites reinforced with 4, 8, 12 and 16 vol% CNT via conventional PM mixing and sintering [45].

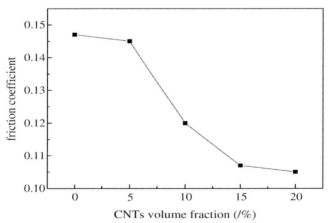

Figure 4.16 Friction coefficient vs carbon nanotube volume content for Al-Mg/MWNT nanocomposites under a load of 30 N. Reproduced with permission from [Chap. 2, Ref. 28]. Copyright © (2007) Elsevier.

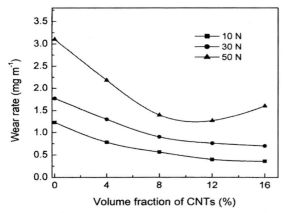

Figure 4.17 Wear rate vs carbon nanotube volume content for Cu/MWNT nanocomposites at different applied loads. Reproduced with permission from [45]. Copyright © (2003) Elsevier.

The CNTs were coated with electrodeless nickel to ensure good wetting between the copper matrix and nanotubes. Pores are formed in such nanocomposites processed by the PM route, ranging from 2.56% for composite filled with 4% CNT to 4.92% for sample with 16 vol% CNT. Figure 4.17 shows the variation of wear rate with nanotube content for the Cu/MWNT nanocomposites subjected to the pin-on-disk test at different applied loads. For applied loads below 30 N, the wear rate generally decreases with increasing nanotube content. At 50 N, the wear rate first decreases with increasing filler content up to 12 vol%, and then increases as the nanotube content reaches 16 vol%. This can be attributed to the high porosity content (4.92%) in the Cu/16 vol% MWNT nanocomposite. Thus, the wear rate of copper nanocomposites depends greatly on the microstructure and applied load under dry sliding conditions.

Dong et al. also studied the sliding wear behavior of Cu/CNT nanocomposites prepared by PM ball milling and sintering (850 °C) [46]. They also fabricated CF/Cu composites in the same way for the purpose of comparison. This is because the Cu/CF composites are widely used for electric brushes and electronic component pedestals applications. They reported that the Cu/CNT nanocomposites exhibit lower friction coefficient and wear loss than those of Cu/CF composites due to the much higher strength of nanotubes. The optimum nanotube content in the nanocomposites for tribological application is 12–15 vol%.

For Cu/MWNT nanocomposites prepared by molecular level mixing and SPS consolidation [Chap. 2, Ref. 87], CNTs are found to disperse homogeneously within the copper matrix; thus MWNT additions enhance the hardness of copper. Consequently, the wear resistance of copper is considerably improved by adding CNTs (Figure 4.18(a) and (b)). Compared with the copper nanocomposites fabricated by PM processing and slid at 30 N (Figure 4.17), Cu/MWNT nanocomposites densified by SPS exhibit much lower wear rates as expected.

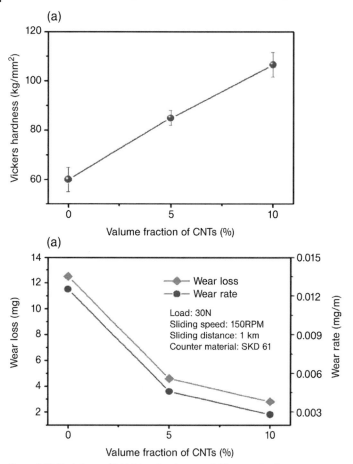

Figure 4.18 Variations of (a) Vickers hardness and (b) wear rate of Cu/MWNT nanocomposites with carbon nanotube volume content under a load of 30 N. Reproduced with permission from [Chap. 2, Ref. 87]. Copyright © (2007) Elsevier.

The tribological behavior of CNT–metal composite coatings will now be considered. Figures 4.19 and 4.20 show the respective variation of friction coefficient and wear volume with applied load for an Ni/MWNT coating subjected to the ball-on-plate test under dry sliding conditions. The coating was deposited onto a carbon steel substrate in a nickel sulfate bath by electrodeless deposition [47]. The nanotube content in the coating as a function of MWNT concentration in the bath is shown in Figure 4.21. It is apparent that the nanotube content in the coating increases with the increase of MWNT concentration in the bath up to $1.1\,\mathrm{g/L^{-1}}$, and thereafter decreases with increasing MWNT concentration. A decrease of nanotube content in the coating at high MWNT concentration is due to the agglomeration of nanotubes in the plating bath. From Figure 4.19, the friction coefficient decreases with increase of the applied load from 10 to 30 N. Further, the coating prepared with

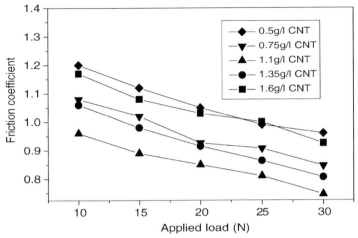

Figure 4.19 Friction coefficient vs applied load for electrodeless deposited Ni/MWNT coatings. Reproduced with permission from [47]. Copyright © (2006) Elsevier.

$1.1\,g/L^{-1}$ CNT exhibits the lowest friction coefficient and wear rate. It is evident that the tribological properties of Ni can be enhanced by increasing the nanotube content in the deposit. A similar beneficial effect of CNTs in improving the wear resistance of Ni-P coating has been reported in the literature [48, 49]. Wang et al. deposited Ni-P/MWNT nanocomposite coating onto a carbon steel substrate in which the MWNTs were shortened by ball milling prior to deposition [49]. The dry sliding pin-on-disk measurement revealed that the friction coefficient of electro-

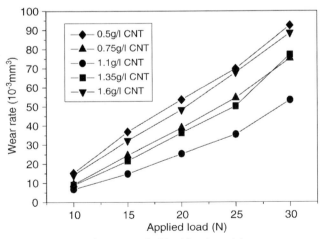

Figure 4.20 Wear volume vs applied load for electrodeless deposited Ni/MWNT coatings. Reproduced with permission from [47]. Copyright © (2006) Elsevier.

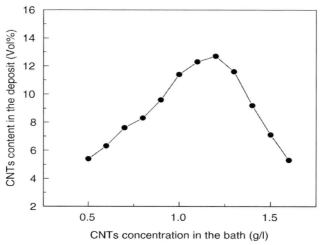

Figure 4.21 Carbon nanotube content in the coatings as a function of MWNT concentration in the plating bath. Reproduced with permission from [47]. Copyright © (2006) Elsevier.

deless plated composite coating decreases with increasing nanotube content due to self-lubrication of nanotubes (Figure 4.22). Further, the wear rate of the composite coating decreased with increasing nanotube content up to 11.2 vol%, but then increased with further increasing filler volume fraction due to clustering of nanotubes.

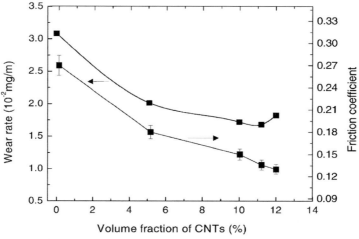

Figure 4.22 Friction coefficient and wear rate as a function of carbon nanotube content for electrodeless deposited Ni-P/MWNT coatings. Reproduced with permission from [49]. Copyright © (2003) Elsevier.

Nomenclature

A	Indentation contact area
β	Dimensionless constant
b	Burgers vector
d	Fiber or particle diameter
E	Elastic modulus
E_c	Composite modulus
E_{CL}	Longitudinal composite modulus
E_{CT}	Transverse composite modulus
E_f	Fiber modulus
E_m	Matrix modulus
E_r	Reduced elastic modulus
F	Frictional force
G	Shear modulus
H	Hardness
k	Wear coefficient
l	Fiber length
L	Sliding distance
M	Taylor factor
μ	Friction coefficient
N	Normal force
P_{max}	Maximum applied load
R	Reinforcement strengthening efficiency
S_c	Stiffness
S	Fiber aspect ratio
S_{eff}	Effective fiber aspect ratio
σ	Yield stress
σ_c	Composite yield stress
σ_m	Matrix yield stress
θ	Misorientation angle
τ	Shear yield stress
υ	Poisson's ratio
V	Wear volume
V_f	Fiber or particle volume fraction
V_m	Matrix volume fraction

References

1 Choi, H.J., Kwon, G.B., Lee, G.Y. and Bae, D.H. (2008) Reinforcement with carbon nanotubes in aluminum matrix composites. *Scripta Materialia*, **59**, 360–363.

2 Gutz, I.A. and Rushchitsky, J.J. (2004) Comparison of mechanical properties and effects in micro- and nanocomposites with carbon fillers (carbon microfibers, graphite microwhiskers and carbon

nanotubes). *Mechanics of Composite Materials*, **40**, 179–190.

3 Bhushan, B. (2005) Nanotribology and nanomechanics. *Wear*, **259**, 1507–1531.

4 Gutz, I.A., Rodger, A.A., Guz, A.N. and Rushchitsky, J.J. (2007) Developing the mechanical models for nanomaterials. *Composites A*, **38**, 1234–1250.

5 Liu, W.K., Park, H.S., Qian, D., Karpov, E.G., Kadowaki, H. and Wagner, G.J. (2006) Bridging scale methods for nanomechanics and materials. *Computer Methods in Applied Mechanics and Engineering*, **195**, 1407–1421.

6 Cox, H.L. (1952) The elasticity and strength of paper and other fibrous materials. *British Journal of Applied Physics*, **3**, 72–79.

7 Nardone, V.C. and Prewo, K.M., (1986) On the strength of discontinuous silicon carbide reinforced aluminum composites. *Scripta Metallurgica*, **20**, 43–48.

8 Nardone, V.C. (1987) Assessment of models used to predict the strength of discontinuous silicon carbide reinforced aluminum alloys. *Scripta Metallurgica*, **21**, 1313–1318.

9 Ryu, H.J., Cha, S.I. and Hong, S.H. (2003) Generalized shear-lag model for load transfer in SiC/Al metal-matrix composites. *Journal of Materials Research*, **18**, 2851–2858.

10 Halpin, J.C. and Kardos, J.L. (1976) The Halpin-Tsai equations: A review. *Polymer Engineering and Science*, **16**, 344–352.

11 Mallick, P.K. (1993) *Fiber-Reinforced Composites: Materials, Manufacturing and Design*, Marcel Dekker, New York, USA, p. 130.

12 An, L., Xu, W., Rajagopalan, S., Wang, C., Wang, H., Fan, Y., Zhang, L., Jiang, D., Kapat, J., Chow, L., Guo, B., Liang, J. and Vaidyanathan, R. (2004) Carbon-nanotube-reinforced polymer-derived ceramic composites. *Advanced Materials*, **16**, 2036–2040.

13 George, R., Kashyap, K.T., Rahul, R. and Yamdagni, S. (2005) Strengthening in carbon nanotube/aluminum (CNT/Al) composites. *Scripta Materialia*, **53**, 1159–1163.

14 Walters, D.A., Ericson, L.M., Casavant, M.J., Liu, J., Colbert, D.T., Smith, K.A. and Smalley, R.E. (1999) Elastic strain of freely suspended single-wall carbon nanotube ropes. *Applied Physics Letters*, **74**, 3803–3805.

15 Kutana, A., Giapis, K.P., Chen, J.Y. and Collier, C.P. (2006) Amplitude response of single-wall carbon nanotube probes during tapping mode atomic force microscopy: Modeling and experiment. *Nano Letters*, **6**, 1669–1673.

16 Oliver, W.C. and Pharr, G.M. (1992) An improved technique for determining hardness and elastic modulus using load and displacement sensing indentation experiments. *Journal of Materials Research*, **7**, 1564–1583.

17 Oliver, W.C. and Pharr, G.M. (2004) Measurement of hardness and elastic modulus by instrumented indentation: Advances in understanding and refinements to methodology. *Journal of Materials Research*, **19**, 3–20.

18 Lee, S.W., Choi, H.J., Kim, Y. and Bae, D.H. (2007) Deformation behavior of nanoparticle/fiber-reinforced nanocrystalline Al-matrix composites. *Materials Science and Engineering A*, **449–451**, 782–785.

19 Goh, C.S., Wei, J., Lee, L.C. and Gupta, M. (2008) Ductility improvement and fatigue studies in Mg-CNT nanocomposites. *Composites Science and Technology*, **68**, 1432–1439.

20 Kim, K.T., Eckert, J., Menzel, S.B., Gemming, T. and Hong, S.H. (2008) Grain refinement assisted strengthening of carbon nanotube reinforced copper matrix nanocomposites. *Applied Physics Letters*, **92**, 121901(1)–121901(3).

21 Zhang, Z. and Chen, D.L. (2006) Consideration of strengthening effect in particulate-reinforced metal matrix nanocomposites: A model for prediction their yield strength. *Scripta Materialia*, **54**, 1321–1326.

22 Orowan, E. (1942) A type of plastic deformation new in metals. *Nature*, **149**, 643–644.

23 Kochs, U.F. (1966) Spacing of dispersed obstacles. *Acta Metallurgica*, **14**, 1629–1631.

24 Hassan, S.F. and Gupta, M. (2006) Effect of type of primary processing on the microstructure, CTE and mechanical properties of magnesium/alumina nanocomposites. *Composite Structures*, **72**, 19–26.

25 Hassan, S.F. and Gupta, M. (2007) Development of nano-Y_2O_3 containing magnesium nanocomposites using solidification processing. *Journal of Alloys and Compounds*, **429**, 176–183.

26 Tun, K.S. and Gupta, M. (2007) Improving mechanical properties of magnesium using nano-yttria reinforcement and microwave assisted powder metallurgy method. *Composites Science and Technology*, **67**, 2657–2664.

27 Archard, F. (1953) Contact and rubbing of flat surfaces. *Journal of Applied Physics*, **24**, 981–988.

28 Suh, N.P. (1977) Overview of delamination theory of wear. *Wear*, **44**, 1–16.

29 Riahi, A.R. and Alpas, A.T. (2001) The role of tribo-layers on the sliding wear behavior of graphitic aluminum matrix composites. *Wear*, **251**, 1396–1407.

30 Akhlaghi, F. and Pelaseyyed, S.A. (2004) Characterization of aluminum/graphite particulate composites synthesized using a novel method termed "*in-situ* powder metallurgy". *Materials Science and Engineering A*, **385**, 258–266.

31 Flores-Zamora, M.I., Estrada-Guel, I., Gonzalez-Hernandez, J., Miki-Yoshida, M. and Martinez-Sanchez, R. (2007) Aluminum-graphite composite produced by mechanical milling and hot extrusion. *Journal of Alloys and Compounds*, **434–435**, 518–521.

32 Kestursatya, M., Kim, J.K. and Rohatgi, P.K. (2003) Wear performance of copper-graphite composite and a leaded copper alloy. *Materials Science and Engineering A*, **339**, 150–158.

33 Chen, X., Zhang, G., Chen, C., Zhou, L., Li, S. and Li, X. (2003) Carbon nanotube composite deposits with high hardness and high wear resistance. *Advanced Engineering Materials*, **5**, 514–518.

34 Wu, S.Q., Zhu, H.G. and Tjong, S.C. (1999) Wear behavior of *in-situ* Al-based composites containing TiB_2, Al_2O_3 and Al_3Ti particles. *Materials Science and Engineering A*, **30**, 243–248.

35 Tjong, S.C. and Lau, K.C. (2000) Dry sliding wear of TiB_2 particle-reinforced aluminum alloy composites. *Materials Science & Technology*, **16**, 99–102.

36 Bhushan, B. and Sundararajan, S. (1998) Micro/nanoscle friction and wear mechanisms of thin films using atomic force and friction force microscopy. *Acta Materialia*, **46**, 3793–3804.

37 Bhushan, B. (2001) Nano- to microscale wear and mechanical characterization using scanning probe microscopy. *Wear*, **251**, 1105–1123.

38 Bhushan, B. (2008) Nanotribology of carbon nanotubes. *Journal of Physics: Condensed Matter*, **20**, 3652141–36521414.

39 Landman, U., Luedtke, W.D., Burham, N.A. and Colton, R.J. (1990) Atomistic mechanisms and dynamics of adhesion, nanoindentation and fracture. *Science*, **248**, 454–461.

40 Buldum, A. and Lu, J.P. (1999) Atomic scale sliding and rolling of carbon nanotubes. *Physical Review Letters*, **83**, 5050–5053.

41 Ni, B. and Sinnott, S.B. (2001) Tribological properties of carbon nanotubes bundles predicted from atomistic simulations. *Surface Science*, **487**, 87–96.

42 Falvo, M.R., Taylor, R.M. II, Helser, A., Chi, V. Jr, Brooks, F.P., Washburn, S. and Superfine, R. (1999) Nanometre-scale rolling and sliding of carbon nanotubes. *Science*, **397**, 236–238.

43 Bhushan, B., Gupta, B.K., Van Cleef, G.W., Capp, C. and Coe, J.V. (1993) Fullerene (C_{60}) films for solid lubrication. *Tribology Transactions*, **36**, 573–580.

44 Zhang, P., Lu, J., Xue, Q. and Liu, W. (2001) Microfrictional behavior of C_{60} particles in

different C$_{60}$ LB films by AFM/FFM. *Langmuir*, **17**, 2143–2145.

45 Chen, W.X., Tu, J.P., Wang, L.Y., Gan, H.Y., Xu, Z.D. and Zhang, X.B. (2003) Tribological application of carbon nanotubes in a metal-based composite coating and composites. *Carbon*, **41**, 215–222.

46 Dong, S.R., Tu, J.P. and Zhang, X.B. (2001) An investigation of the sliding wear behavior of Cu-matrix composite reinforced by carbon nanotubes. *Materials Science and Engineering A*, **313**, 83–87.

47 Chen, X.H., Chen, C.S., Xiao, H.N., Liu, H.B., Zhou, L.P., Li, S.L. and Zhang, G. (2006) Dry sliding and wear characteristics of nickel/carbon nanotube electrodeless composite deposits. *Tribology International*, **39**, 22–28.

48 Li, Z.H., Wang, X.Q., Wang, M., Wang, F.F. and Ge, H.L. (2006) Preparation and tribological properties of the carbon nanotubes-Ni-P composite coating. *Tribology International*, **39**, 953–957.

49 Wang, L.Y., Tu, J.P., Chen, W.X., Wang, Y.C., Liu, X.K., Olk, C., Cheng, D.H. and Zhang, X.B. (2003) Friction and wear behavior of electrodeless Ni-based CNT composite coating. *Wear*, **254**, 1289–1293.

5
Carbon Nanotube–Ceramic Nanocomposites

5.1
Overview

The development of advanced technologies in the aeronautics, space and energy sectors requires high performance materials with excellent mechanical properties, high thermal conductivity and good wear resistance. Metallic composites can meet these requirements but they suffer from corrosion and oxidation upon exposure to severe aggressive environments. Further, composite materials based on aluminum have a low melting point (~643 °C) that precludes their use as structural materials at elevated temperatures. Ceramic-based materials such as zirconia (ZrO_2), alumina (Al_2O_3), silicon carbide (SiC), silicon nitride (Si_3N_4) and titanium carbide (TiC) have been used in industrial sectors at high temperatures due to their intrinsic thermal stability, good corrosion resistance, high temperature mechanical strength and low density. However, ceramics are known to exhibit low fracture toughness since plastic deformation in ceramics is very limited. Several approaches have been adopted to improve the fracture toughness of ceramics. These include transformation toughening, ductile-phase toughening and reinforcement toughening [1] (Figure 5.1).

Transformation toughening involves the occurrence of phase transformation in zirconia-based ceramics to arrest the propagation of cracks. Pure zirconia exhibits three different crystalline structures: monoclinic (room temperature to 1170 °C), tetragonal (1170–2370 °C) and cubic (>2370 °C). Several stabilizers or dopants are known to stabilize the tetragonal and cubic phases at room temperature in the metastable state [2, 3]. Partial stabilization enables retention of the metastable tetragonal phase of zirconia at ambient temperature by adding appropriate dopants such as MgO, CaO and Y_2O_3 (designated as M-TZP, C-TZP and Y-TZP). Under external stress loading, phase transformation from the tetragonal to the monoclinic phase occurs in the stress field around the crack tip [4]. This stress-induced tetragonal to monoclinic transformation is commonly referred to as "martensitic transformation" and somewhat similar to that in the carbon steels. Accompanying this transformation is a large increase in volume expansion. The resulting strain tends to relieve the stress field across the crack tip, thereby facilitating absorption of large fracture energy.

Carbon Nanotube Reinforced Composites: Metal and Ceramic Matrices. Sie Chin Tjong
Copyright © 2009 WILEY-VCH Verlag GmbH & Co. KGaA, Weinheim
ISBN: 978-3-527-40892-4

(a)
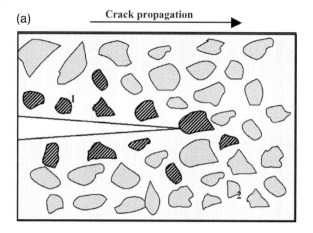

(b)
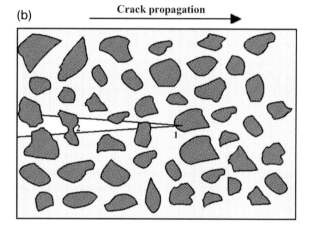

(c)
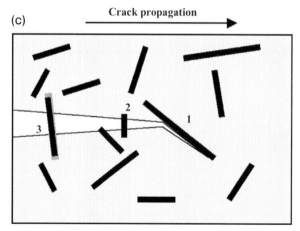

Ductile phase toughening involves the incorporation of a ductile metal phase into brittle ceramics to facilitate yielding of the metal particle and blunting of the propagating cracks. In this regard, energy is dissipated through the deformation of the ductile phase and crack blunting at the ductile particle. In most cases, bridging ligaments along the crack profiles are also produced [5–7]. In general, large amounts of metallic particles (30–70%) are needed to improve the fracture toughness of ceramics [8]. Addition of a large amount of metal microparticles deteriorates the sinterability of ceramics significantly. Further, failure of such MMCs originates from cavities in large metal inclusions [9].

Fracture toughness of ceramics can also be improved by the addition of ceramic reinforcements in the forms of particulates, whiskers, and fibers to form ceramic-matrix composites (CMCs) [10–14]. The reinforcing effect of fibers is much higher that that of particulates and whiskers. Continuous silicon carbide and carbon fibers have been widely used to reinforce ceramics [15–17]. The toughening mechanisms of fiber-reinforced CMCs are mainly attributed to the crack deflection at the fiber–matrix interface, crack bridging and fiber pull-out. It has been demonstrated that weak fiber–matrix interfacial bonding facilitates the fiber pull-out toughening mechanism to operate. This is because strong interfacial bonding allows the crack to propagate straight through the fibers, resulting in low fracture toughness [15]. Continuous fiber-reinforced CMCs are generally fabricated by chemical vapor infiltration or polymer infiltration and pyrolysis (PIP). However, these processing techniques are costly and time consuming.

5.2
Importance of Ceramic-Matrix Nanocomposites

In the past few years, considerable attention has been paid to the development of nanocrystalline ceramics with improved mechanical strength and stiffness, and enhanced wear resistance [18–20]. Decreasing the grain size of ceramics to the submicrometer/nanometer scale leads to a marked increase in hardness and fracture strength. However, nanocrystalline ceramics generally display worse fracture toughness than their microcrystalline counterparts [18]. The toughness of nanoceramics can be enhanced by adding second phase reinforcements.

Grain size refinement in ceramics and their composites can yield superplasticity at high strain rates [21, 22]. Superplasticity is a flow process in which crystalline materials exhibit very high tensile ductility or elongation prior to final failure at high

Figure 5.1 Schematic illustrations of three toughening mechanisms in ceramics: (a) phase transformation toughening showing transformed monoclinic zirconia (1) from metastable tetragonal zirconia (2); (b) ductile phase toughening through (1) ductile phase deformation or crack blunting and (2) crack bridging; (c) fiber toughening showing (1) crack deflection, (2) crack bridging and (3) fiber pull-out. Reproduced with permission from [1]. Copyright © (2005) Elsevier.

temperatures. Superplastic deformation is of technological interest because it allows lower processing temperature and time, and enables near net shape forming of ceramic products. The ability to prevent premature failure of ceramics at high strain rates can have a large impact on the production processes.

From the concept of new materials design, Niihara classified the nanocomposites into four types based on their microstructural aspects that is, intra, inter, intra/inter and nano-nano [23]. The first three types describe the dispersion of second-phase nanoparticles within matrix micrograins, at the grain boundaries or both. These composites are commonly known as ceramic nanocomposites but are more precisely referred to as micro-nano composites. For the nano-nano type, second-phase nanoparticles are embedded within nanograined matrix. Most of the nanocomposites reported in the literature belong to the micro-nano type [24–33]. Less information is available for the nano-nano ceramic composites [34–36]. Mukherjee later proposed a new classification of ceramic composites: nano-nano, nano-micro, nano-fiber and nano-nanolayer [35]. In this classification, the matrix phase is continuous nanocrystalline grains of less than 100 nm while the second phase could be in the form of nanoparticles, microparticles, fibers (or whiskers) or a grain boundary nanolayer (Figure 5.2).

Niihara *et al.* reported that the incorporation of 5% SiC nanoparticles into microcrystalline alumina increased the flexural strength from 350 MPa to 1.1 GPa, and further annealing increased the strength to more than 1.5 GPa. The fracture toughness improved from 3.5 MPa m$^{1/2}$ to 4.8 MPa m$^{1/2}$. Addition of SiC nanoparticles to alumina matrix also produced a change in the fracture mode of the matrix from predominantly intergranular to transgranular mode [23]. Transgranular failure

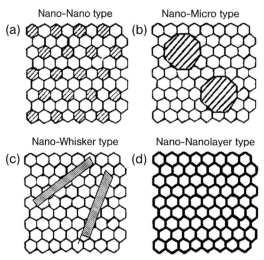

Figure 5.2 Mukherjee's classification of nanocomposite types, based on nanosized matrix grains and second-phase particle of different morphologies and sizes. Reproduced with permission from [35]. Copyright © (2001) Elsevier.

has also been reported by other researchers for the micro-nano ceramic composites [25–28]. The switch from intergranular to transgranular failure is a good indicator for the strength and toughness improvements in ceramic nanocomposites. Second-phase nanoparticles enhance grain boundary properties of ceramics by inducing residual stress as a result of thermal expansion mismatch between the matrix and particles. This internal stress causes an advancing crack to propagate in a tortuous manner rather than a catastrophic straight path. Moreover, second-phase nanoparticle additions also lead to the grain refinement of the ceramic matrix [37]. Reducing grain size promotes the mechanical strength of crystalline materials according to the Hall–Petch relation. In some cases, second-phase nanoparticles also yield elongation of matrix grains. The elongated matrix grains with high aspect ratio increase the fracture toughness by dissipating the fracture energy through crack deflection and bridging mechanisms [30, 31]. Table 5.1 lists the properties of typical micro-nano and nano-nano ceramic composites.

Carbon nanotubes (CNTs) with high aspect ratio, extraordinary mechanical strength and stiffness, excellent thermal and electrical conductivity are attractive nanofillers which produce high-performance ceramic composites with multifunctional properties. The reinforcing effect of CNTs with high aspect ratio is considered to be analogous to that of continuous or short-fiber-reinforcement. The superior flexibility of CNTs is very effective in improving the fracture toughness of brittle ceramics. This is accomplished by means of crack deflection at the CNT–matrix interface, crack-bridging and CNT pull-out mechanisms. Recently, Huang et al. reported that SWNTs exhibit superplastic deformation with an apparent elongation of 280% at high temperatures [38]. This finding shows the potential application of CNTs

Table 5.1 Mechanical properties of typical micro-nano and nano-nano ceramic composites.

Investigators	Materials	Bending strength (MPa)	Fracture toughness (MPa m$^{1/2}$)
	Micro-nano type:		
Anya Ref [25]	Monolithic Al$_2$O$_3$ (3.5 µm)	430	3.2
	Al$_2$O$_3$(4.0 µm)/5 vol% SiC (200 nm)	646	4.6
	Al$_2$O$_3$(2.9 µm)/10 vol% SiC (200 nm)	560	5.2
	Al$_2$O$_3$(2.6 µm)/15 vol% SiC (200 nm)	549	5.5
Gao et al. Ref [26]	Monolithic Al$_2$O$_3$	350	3.5
	Al$_2$O$_3$(1.8 µm)/5 vol% SiC (70 nm)	1000	∼ 4.0
Zhu et al. Ref [30]	Monolithic Al$_2$O$_3$	380	3.9
	Al$_2$O$_3$/15vol%Si$_3$N$_4$ (200 nm, inter; 80 nm, intra)	820	6.0
Isobe et al. Ref [33]	Monolithic Al$_2$O$_3$ (1.95 µm)	450	3.7
	Al$_2$O$_3$ (0.91 µm)/2.8 vol% Ni (150 nm, inter; 300 nm, intra)	766	4.6
	Nano-nano type:		
Bhaduri Ref [34]	Al$_2$O$_3$ (44 nm)/10 vol% ZrO$_2$(12.7 nm)	—	8.38

as reinforcing fillers for CMCs with improved ductility. Thus, such ceramic-CNT nanocomposite could possess superplastic deformability. Peigney et al. investigated the extruding characteristics of metal oxide-CNT nanocomposites at high temperatures. They indicated that superplastic forming of nanocomposites is made easier by adding CNTs [39]. All these attractive and unique properties of CNTs enable materials scientists to create novel strong and tough ceramic nanocomposites. Moreover, the electrical and thermal conductivities of ceramics can be improved markedly by adding nanotubes. The electrical conductivity of alumina-CNT composites can reach up to twelve orders of magnitude higher than their monolithic counterpart. Recent study has shown that the thermal conductivity of alumina-CNT nanocomposites exhibits anisotropic behavior [40]. The nanocomposites conduct heat in one direction, along the alignment of the nanotube axial direction, but reflect heat at right angles to the nanotubes. This anisotropic thermal behavior makes alumina-CNT nanocomposites potential materials for application as thermal barrier layers in microelectronic devices, microwave devices, solid fuel cells, chemical sensors, and so on [40].

5.3
Preparation of Ceramic-CNT Nanocomposites

Despite the fact that the CNTs exhibit remarkable mechanical properties, the reinforcing effect of CNTs in ceramics is far below our expectation. The problems arise from inhomogeneous dispersion of CNTs within the ceramic matrix, inadequate densification of the composites and poor wetting behavior between CNTs and the matrix. All these issues are closely related to the fabrication processes for making ceramic-CNT nanocomposites. As recognized, CNTs are hard to disperse in ceramics. They tend to form clusters caused by van der Waals force interactions. Such clustering produces a negative effect on the physical and mechanical properties of the resulting composites. Individual nanotubes within clusters may slide against each other during mechanical deformation, thereby decreasing the load transfer efficiency. Furthermore, toughening of the ceramic matrix is difficult to achieve if the CNTs agglomerate into clusters. Carbon nanotube clusters minimize crack bridging and pull-out effects greatly. Therefore, homogeneous dispersion of CNTs in the ceramic matrix is a prerequisite of achieving the desired mechanical properties.

In most cases, ceramic-CNT composites were prepared by conventional powder mixing and sintering techniques such as pressureless sintering, hot pressing and hot isostatic pressing (HIP). The mechanical properties of ceramic-CNT nanocomposites sintered from these techniques show modest improvement or even deterioration. This is because conventional sintering methods require lengthy treatment at high temperatures to densify green compacts. Such a high temperature environment causes oxidation and deterioration in the properties of CNTs. For example, Ma et al. prepared the SiC/10 vol% MWNT nanocomposite by ultrasonically mixing SiC nanoparticles with CNTs and hot pressing at 2000 °C. They reported a modest improvement in bending strength and toughness of nanocomposite over a monolithic SiC specimen fabricated under similar processing conditions. A reasonable relative density of ~95% has been achieved by hot pressing [41]. Peigney's group fabricated Fe-Al_2O_3/CNT nanocomposites by hot pressing composite powder mixtures at 1335–1535 °C [42]. The relative density of Fe-Al_2O_3/CNT nanocomposites is lower that that of Fe-Al_2O_3 composites. Carbon nanotubes tend to inhibit the densification of nanocomposites, particularly for materials with higher filler loadings. Bundles of CNTs are found in hot-pressed nanocomposites. The fracture strength of Fe-Al_2O_3/CNT nanocomposites is only marginally higher than that of monolithic alumina, but lower than that of the Fe-Al_2O_3 composites. Further, the fracture toughness of Fe-Al_2O_3/CNT nanocomposites is lower than that of Al_2O_3.

Spark plasma sintering (SPS) is recognized as an effective process for achieving higher densification of ceramics at a relatively lower temperature with short holding time. Accordingly, several researchers have used the SPS method to consolidate ceramic-CNT nanocomposites [43–45]. For instance, Balazsi et al. compared the effects of HIP and SPS treatments on the microstructural and mechanical properties of Si_3N_4/MWNT nanocomposites [43]. Large differences in the properties of composites prepared by these two sintering techniques have been found. Fully dense nanocomposites with improved mechanical properties can be achieved using the SPS method. In contrast, HIP-treated composites exhibit a partially dense structure with coarser grains.

Ceramic-CNT interfacial behavior is another key factor in controlling the mechanical and physical properties of the nanocomposites. Poor compatibility and wettability between CNTs and ceramics result in weak interfacial strength. The interface cannot transfer the applied load effectively under low interfacial shear strength. Thus stronger bonds between CNTs and the ceramic matrix are needed for effective load transfer to occur. Interfacial bonding in ceramic-CNT nanocomposites is less well understood compared with that of polymer-CNT nanocomposites. Both covalent and non-covalent bonding between the CNT and polymer can be established by forming surface functional groups on nanotubes by acid treatment, fluorination and functionalization [46].

One possible approach to improve the wettability between CNTs and inorganic ceramic host is to coat CNT surfaces with proper surfactants [47–50]. However, surfactants may introduce undesirable impurities that can affect the sintering process and resulting ceramic-CNT composite properties. It has been reported that molecular level mixing [51], aqueous colloid [52–55] and polymer-derived ceramics [56] processes can yield homogeneous dispersion of CNTs and strong interfacial strength in the ceramic-CNT nanocomposites. For example, Fan et al. reported that the fracture toughness of the alumina/SWNT nanocomposites is twice that of monolithic alumina. They attributed this to the strong interfacial CNT-alumina bonding obtained by heterocoagulation [54, 55].

In general, strong interfacial bonding facilitates effective load transfer effect, but it prevents nanotube pull-out toughening from occurring. Weak interfacial bonding favors CNTs pull-out but fails to strengthen the ceramic matrix. Thus, a balance must be maintained between nanotube pull-out and strengthening mechanisms. It is well recognized that improved toughness of continuous fiber-reinforced ceramic composites is obtained under moderate fiber–ceramic interfacial bonding. In this regard, suitable (neither too strong nor too weak) ceramic-nanotube interfacial bonding is needed to ensure effective load transfer, and to enhance the toughness and strength of ceramic-CNT nanocomposites.

5.4
Oxide-Based Nanocomposites

5.4.1
Alumina Matrix

5.4.1.1 Hot Pressing/Extrusion

Alumina ceramics exhibit high hardness and good wear resistance, but possess poor flexural strength and fracture toughness. Many attempts have been made to improve their mechanical properties by adding CNTs. Homogenous dispersion of CNTs in alumina is difficult to obtain. For most Al_2O_3/CNT nanocomposites, CNTs were synthesized independently, and then introduced into micro- or nano alumina powders under sonication to form nanocomposites [57]. This may lead to poor dispersion of nanotubes in the alumina matrix. A growing interest has been directed toward the synthesis of Al_2O_3/CNT nanocomposites by the *in situ* reaction method. This can be achieved by exposing transition metal/metal oxide catalysts to reactive gases at high temperatures. The advantage of this technique is that the CNTs formed *in situ* are directly incorporated into the alumina matrix during the synthetic process. For instance, Peigney *et al.* used a catalysis method for the *in situ* production of composite powders. The Fe-Al_2O_3/CNT nanocomposite powders were synthesized by nucleating double-walled CNTs or SWNTs *in situ* on metal oxide solid solutions [42, 58]. The process involved the initial formation of $Al_{2-2x}Fe_{2x}O_3$ solid solutions from the decomposition and calcination of corresponding mixed oxalates [42, 58]. Selective reduction of such solid solutions in a H_2-CH_4 atmosphere produced Fe nanoparticles that acting as active catalysts for the *in situ* nucleation and growth of CNTs. Synthesized composite powders were then subjected to hot pressing. The quality and quantity of the CNTs depend on the Fe content (5–20 wt%) in the oxide solid solution and the reduction temperature (900 or 1000 °C). They also synthesized $MgAl_2O_4$ and MgO matrix composites reinforced with CNTs. These metal-oxide/CNT composite powders were prepared by reducing oxide solid solutions based on α-Al_2O_3, MgO or $MgAl_2O_4$, and Fe, Co or FeCo alloy in a H_2-CH_4 atmosphere [59, 60].

Figure 5.3(a) and (b) show SEM micrographs of synthesized Fe-Al_2O_3/4.8 wt% CNT and Fe-Al_2O_3/5.7 wt% CNT composite powders. Carbon nanotubes are arranged in long web-like bundles and homogeneously dispersed between the metal-oxide grains. These composite powders were hot-pressed at 1500 °C. The relative density values of consolidated Fe-Al_2O_3/4.8 wt% CNT and Fe-Al_2O_3/5.7 wt% CNT composites are 88.7 and 87.3%, respectively. A fraction of CNTs is destroyed during

5.4 Oxide-Based Nanocomposites

Figure 5.3 SEM micrographs of (a) Fe-Al$_2$O$_3$/4.8 wt% CNT and (b) Fe-Al$_2$O$_3$/5.7 wt% CNT composite powders. Reproduced with permission from [59]. Copyright © (2000) Elsevier.

hot pressing at 1500 °C, leading to the formation of disordered graphene sheets at grain boundary junctions (Figure 5.4(a) and (b)). Obviously, dense alumina-CNT nanocomposite cannot be achieved by hot-pressing *in situ* composite powders at high temperatures. The low density of the resulting nanocomposites affects their mechanical strength and fracture toughness dramatically.

In another study, Peigney *et al.* used high-temperature extrusion to consolidate the *in situ* composite powders [39]. They demonstrated that the CNTs can withstand extreme shear stresses during extrusion. The densification of the extruded composite

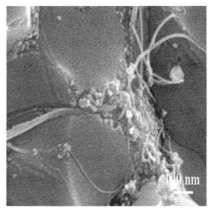 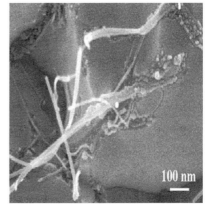

Figure 5.4 SEM images of the fracture surface of hot-pressed Fe-Al$_2$O$_3$/4.8 wt% CNT nanocomposite showing formation of disordered graphene sheets at grain junctions. Reproduced with permission from [59]. Copyright © (2000) Elsevier.

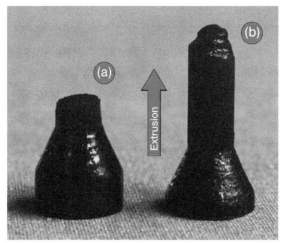

Figure 5.5 Photographs of Fe-Al$_2$O$_3$/CNT composite (a) at the beginning of the extrusion process and (b) after the extrusion is completed. Reproduced with permission from [39]. Copyright © (2002) Elsevier.

is ~90%, which is slightly higher that that of hot-pressed composite. Figure 5.5 shows a photograph of Fe-Al$_2$O$_3$/CNT composite extruded at 1500 °C. Carbon nanotubes inhibit the grain growth of ceramic matrix and promote superplastic forming of the composite during extrusion. Moreover, CNTs tend to align preferentially along the extrusion direction (Figure 5.6(a) and (b)).

In addition to the *in situ* formation of ceramic-CNT powder mixtures, other potential techniques for homogeneous dispersion of CNTs in alumina matrix have been sought. The fabrication methods should allow easier dispersion of nanofillers into ceramic powders. These include aqueous colloidal processing and molecular level mixing. Aqueous colloidal processing consists of several step procedures: (i) suspension preparation, (ii) removal of the solvent phase, (iii) consolidation into desired

 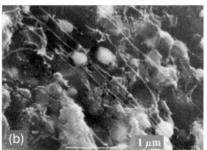

Figure 5.6 SEM images showing the fracture surfaces of extruded (a) Fe-Al$_2$O$_3$/CNT and (b) FeCo-MgAl$_2$O$_4$/CNT nanocomposites. Note the alignment of CNTs. Reproduced with permission from [39]. Copyright © (2002) Elsevier.

component shape [53–55, 61–63]. A stable colloid with well dispersed particles produces a dense and homogeneous powder. The interparticle colloidal forces can be manipulated by adjusting the pH or by adding a dispersant. The stability of aqueous colloidal suspension can be controlled by creating like charges of sufficient magnitude on the surfaces of ceramic particles, known as electrostatic stabilization. Alternatively, ionic polymer dispersant is added to the suspension such that polymeric chains adsorb on ceramic particle surfaces creating steric repulsion. Electrosteric stabilization that combines both ionic and polymer dispersant mechanisms is more effective for obtaining a well-dispersed suspension [64, 65].

As CNTs and alumina particles show extensive agglomeration, they must disperse independently in organic suspensions. When two sols of opposite sign are mixed, mutual coagulation takes place. Under a properly selected pH range, alumina particles adsorb onto CNT surfaces through electrostatic interaction. Recently, Sun et al. fabricated alumina-CNT nanocomposites by means of colloidal processing [61]. Carbon nanotubes were treated with ammonia gas at 600 °C for 3 h, and then dispersed in cationic type polyethyleneamine (PEI) solution. The treated CNTs exhibit positive charge in a wide pH range on the basis of zeta potential measurement. Zeta potential is defined as the electrokinetic potential of particulate dispersions associated with the magnitude of electrical charge at the double layer of colloidal systems. It is commonly used to measure the magnitude of attraction or repulsion between dispersed particles [66]. Alumina nanoparticles (30 nm) were dispersed independently in anionic type poly(acrylic acid) (PAA) solution, yielding electronegative charges on particle surfaces. Sodium hydroxide was used to adjust the pH of suspension solution. The alumina suspension with PAA was dripped into the CNT suspension containing PEI under sonication. The pH value was kept at 8 for the coating of alumina on CNTs.

In another study, pristine CNTs were dispersed in anionic type sodium dodecyl sulfate (SDS) solution, forming electronegative charges on nanotube surfaces [62]. SDS is recognized as an effective dispersing agent for CNTs [67]. Figure 5.7 shows

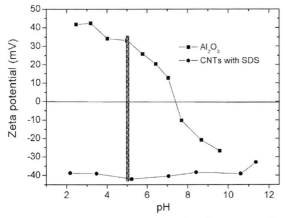

Figure 5.7 Zeta potential as a function of pH for alumina and CNTs in the presence of SDS. Reproduced with permission from [62]. Copyright © (2003) Elsevier.

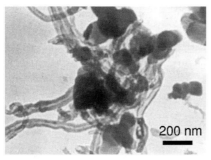

Figure 5.8 TEM image of CNTs coated with alumina in the presence of SDS. Reproduced with permission from [62]. Copyright © (2003) Elsevier.

zeta potential vs pH plot for pristine CNT in the presence of SDS and alumina in 1 mM NaCl solution. The pH value was adjusted to 5 for coating alumina particles on CNTs. The zeta values of alumina and SDS-treated CNTs at pH 5 are about 35 and −40 mV. A very dilute alumina suspension was added to the CNT suspension. Carbon nanotube surfaces were then coated with alumina particles due to the electrostatic attraction (Figure 5.8). Since then, other researchers have employed colloidal processing to prepare alumina-CNT nanocomposites using SDS dispersant [54, 55, 68].

Figure 5.9(a) shows SEM micrograph of hot-pressed Al_2O_3/MWNT nanocomposite prepared by heterocoagulation using a SDS dispersant. MWNTs are found to disperse within alumina grains and at the grain boundaries. The SEM image of the polished nanocomposite specimen subjected to ion milling is shown in Figure 5.9(b). This micrograph clearly reveals homogenous dispersion of MWNTs within the matrix of the nanocomposite. The dispersion of MWNTs within the grains and grain boundaries of the alumina matrix can be readily seen in the transmission electron microscopyTEM micrographs as shown in Figure 5.9(c) and (d).

5.4.1.2 Spark Plasma Sintering

Murkerjee's group prepared Al_2O_3/5.7 vol% SWNT and Al_2O_3/10 vol% SWNT nanocomposites by blending and ball milling SWNTs with alumina nanopowders, followed by spark plasma sintering at 1150 °C for 2–3 min [44]. Pure alumina and Al_2O_3/5.7 vol% SWNT and Al_2O_3/10 vol% SWNT nanocomposites can be consolidated to full density of 100% at 1150 °C for 3 min (Table 5.2). The matrix grain size of the Al_2O_3/5.7 vol% SWNT and Al_2O_3/10 vol% SWNT nanocomposites is considerably smaller than that of pure alumina. Thus, SWNTs are very effective to retard the grain growth of alumina during SPS. Moreover, CNTs tend to form a network structure at grain boundaries (Figure 5.10). Generally, sintering time and temperature have a large influence on the density of resulting nanocomposites. For example, the relative density of the Al_2O_3/10 vol% SWNT nanocomposite drops to 95.2% by reducing the sintering time to 2 min at 1150 °C. At 1100 °C for 3 min, the relative density of the Al_2O_3/10 vol% SWNT nanocomposite reduces to 85.8%.

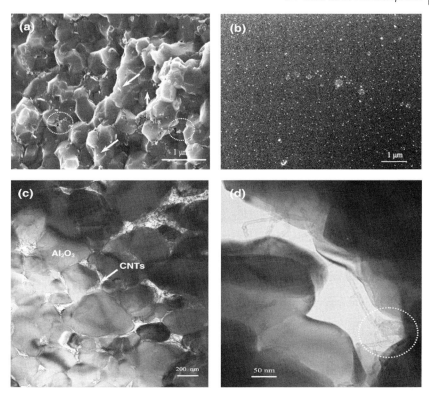

Figure 5.9 SEM images of (a) fractured and (b) polished/ion-milled Al$_2$O$_3$/3 vol% MWNT nanocomposite. White dots in (b) are MWNTs. The circles and arrows indicate CNTs at the grain boundaries and within the grains, respectively. (c) and (d) are TEM images of polished/ion-milled nanocomposite showing the dispersion of nanotubes within the grains and at grain boundaries. Reproduced with permission from [68]. Copyright © (2008) Elsevier.

Table 5.2 Spark plasma sintering conditions and the resultant properties of monolithic alumina and Al$_2$O$_3$/SWNT nanocomposites.

Materials	SPS conditions	Relative density (%)	Grain size (nm)	Hardness (GPa)	Fracture toughness (MPa m$^{1/2}$)
Al$_2$O$_3$	1150 °C, 3 min	100	349	20.3	3.3
Al$_2$O$_3$/5.7 vol% SWNT	1150 °C, 3 min	100	~200	20.0	7.9
Al$_2$O$_3$/10 vol% SWNT	1150 °C, 3 min	100	~200	16.1	9.7
Al$_2$O$_3$/10 vol% SWNT	1150 °C, 2 min	95.2	156	9.30	8.1
Al$_2$O$_3$/10 vol% SWNT	1100 °C, 3 min	85.8	150	4.40	6.4

Reproduced with permission from [44]. Copyright © (2003) Nature Publishing Group.

Figure 5.10 SEM fractograph of Al$_2$O$_3$/5.7 vol% SWNT nanocomposite showing the nanotube network encompassing alumina grains. Reproduced with permission from [44]. Copyright © (2003) Nature Publishing Group.

Murkerjee and coworkers further investigated the effect of SPS temperatures on the microstructure and mechanical properties of the Al$_2$O$_3$/10 vol% SWNT nanocomposite. The nanocomposite specimens were sintered at 1150, 1350 and 1550 °C for 3 min [69, 70]. Figure 5.11 shows the Raman spectra of the Al$_2$O$_3$/10 vol% SWNT nanocomposite sintered at 1150, 1350 and 1550 °C for 3 min. The Raman spectrum of pure SWNT is also shown for the purpose of comparison. For SWNT, the peak at 1350 cm^{-1} corresponds to the D-band associated with the presence of disorder or small sp^2 carbon crystallite. The peak at 1595 cm^{-1} is assigned to the G-band resulting from the vibration mode (C–C stretching) of sp^2-bonded carbon atoms. A shoulder at ~1560–1575 cm^{-1} is due to the overlapping of electrons within graphene layers upon rolling the layers into nanotubes. From Figure 5.11, the

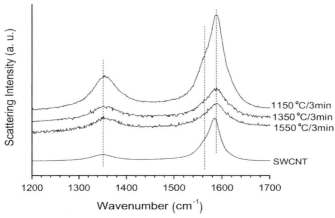

Figure 5.11 Pulsed laser Raman spectra of Al$_2$O$_3$/10 vol% SWNT nanocomposite spark plasma sintered at 1150, 1350 and 1550 °C for 3 min. The Raman spectrum of pristine SWNT is also shown for comparison. Reproduced with permission from [70]. Copyright © (2007) Elsevier.

structure of SWNT of the nanocomposite sintered at 1150 °C is still preserved as evidenced by the presence of G-peak at ~1595 cm^{-1} with a shoulder. The shoulder disappears at higher sintering temperatures of 1350 and 1550 °C, indicating the loss of CNT structure. Moreover, the peak intensity of D-band relative to the G-band decreases with increasing sintering temperatures, implying an increase in the graphite content. In other words, SWNTs convert to graphite structure at temperatures above 1150 °C. The relative density values of the nanocomposite sintered at 1150, 1350 and 1550 °C are 100, 98.9 and 97.8%.

Apparently, the best combination of microstructure and mechanical properties in alumina-CNT nanocomposites can be attained through heterocoagulation and SPS at lower temperature [60]. Poyato *et al.* indicated that sintering the colloidally dispersed composite powder at high temperature (1550 °C) leads to the conversion of CNTs into disordered graphite, diamond and carbon nano-onions [53]. The disintegration of CNTs at high temperatures degrades the mechanical properties of resulting nanocomposites considerably. Up till now, some workers have been quite unaware of the damage caused by high temperature ($\geq$1250 °C) exposure to CNTs. They have still conducted SPS of alumina-CNT nanocomposites at 1500 °C [71–73].

The colloidally formed composite powders, as mentioned above, contain organic surfactant or dispersant [60, 61]. More recently, Estili and Kawasaki employed the heterocoagulation technique to prepare the alumina/MWNT nanocomposites without using any organic dispersants for CNTs and alumina powders [74]. Instead, they dispersed acid purified MWNTs and α-alumina powder (150 nm) in water independently. Carboxylic and hydroxyl functional groups formed on acid-treated MWNTs allow them to be dispersed easily in polar water solvent. A CNT suspension was added to the alumina suspension under a controlled pH of 4.4 (Figure 5.12(a)). The CNT surfaces became coated with alumina particles due to the electrostatic attraction between positively charged alumina particles and negatively charged CNTs at pH 4.4 (Figure 5.12(b) and (c)).

Hong and coworkers demonstrated that a molecular level mixing technique can be used to disperse MWNTs homogeneously in an alumina matrix [72]. Acid-functionalized MWNTs were ultrasonically dispersed in distilled water. Al(NO$_3$)$_3$·9H$_2$O was added to a CNT suspension followed by sonication and vaporization. The retained powders were oxidized at 350 °C for 6 h in air atmosphere, and finally consolidated by SPS at 1500 °C for 5 min. The essence of this process is that MWNTs and metal ions are mixed homogeneously in an aqueous solution at the molecular level. Chemical bonding between MWNT and amorphous Al$_2$O$_3$ was developed during the calcination process (Figure 5.13(a) and (b)). However, homogeneous dispersion of nanotubes is only found at low nanotube loading of 1 vol%. Agglomeration of CNTs occurs in the composite at very low filler loading of 1.8 vol%.

As mentioned before, the nanocomposites can be classified into intra, inter, intra/inter or nano-nano depending on the dispersion of second-phase reinforcement in materials [23]. Recent advances in nanotechnology have made possible the design and development of ceramic nanocomposites reinforced with hybrid nanofillers. Very recently, Ahmad and Pan developed novel hybrid alumina nanocomposites

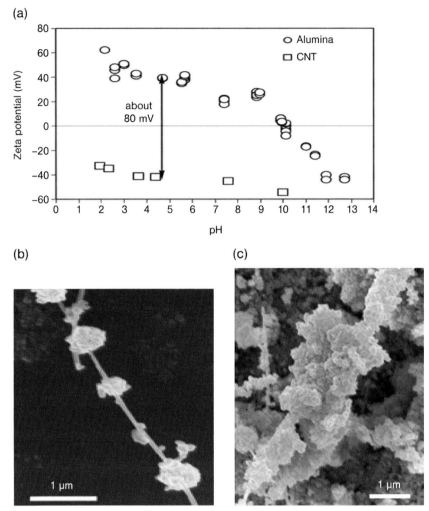

Figure 5.12 (a) Zeta potential values of acid treated MWNTs and alumina at different pH. TEM micrographs showing (b) partial and (c) complete coverage of nanotubes with alumina particles. Reproduced with permission from [74]. Copyright © (2008) Elsevier.

reinforced with two materials, that is, MWNTs (5, 7 and 10 vol%) and SiC nanoparticles (1 vol%) [75]. The fabrication process consisted of blending MWNTs, SiC and alumina in ethanol ultrasonically followed by ball-milling and drying. Dried composite powders were spark plasma sintered at 1550 °C under a pressure of 50 MPa. For these hybrids, SiC nanoparticles are designed to disperse within the grains and grain boundaries of alumina (intra/inter type), and MWNTs are located at grain boundaries as intergranular reinforcements (Figure 5.14). Therefore, hybrid

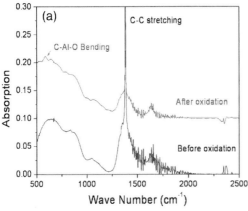
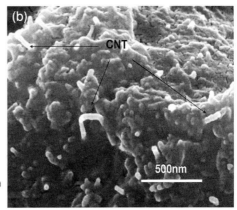

Figure 5.13 (a) FTIR of amorphous Al$_2$O$_3$/MWNT composite powders before and after calcination process; (b) SEM micrograph of the composite powder surface showing homogeneous dispersion and incorporation of CNTs into amorphous Al$_2$O$_3$ powder. Reproduced with permission from [72]. Copyright © (2005) Elsevier.

nanocomposites inherit the beneficial properties of reinforcing MWNTs and SiC nanoparticles, leading to improvement in fracture toughness, bending strength and hardness of the hybrids. Figure 5.15(a) shows a SEM fractograph of the Al$_2$O$_3$/(1 vol% MWNT + 1 vol% SiC) hybrid nanocomposite. MWNTs are mainly dispersed along grain boundaries of alumina, forming a network microstructure. SiC nanoparticles can only be seen in the TEM micrograph as shown in Figure 5.15(b). The fracture mode of hybrid is mixed intergranular and transgranular mode.

5.4.1.3 Plasma Spraying

Very recently, Agarwal's group employed the plasma spraying process to form Al$_2$O$_3$/ 4 wt% MWNT and Al$_2$O$_3$/8 wt% MWNT nanocomposite coatings [76, 77]. They indicated that the Al$_2$O$_3$ nanopowder (150 nm) cannot be plasma sprayed because it has a high tendency to form clogs at the plasma gun nozzle. To facilitate the flowability of the nanopowder, a spray-drying pretreatment for nanopowder was needed. Spray-drying involves the dispersion of nanopowder in an aqueous organic binder to form a slurry followed by atomization. Spray-drying of Al$_2$O$_3$ nanopowder results in the formation of spherical agglomerates of ∼35 μm (designated as A-SD). A-SD powder was then blended with 4 wt% MWNT in a jar mill to form the powder feedstock (designated as A4C-B) for plasma spraying (Figure 5.16). In another approach, mixed alumina nanopowder and MWNT of different contents were spray-dried to form the A4C-SD or A8C-SD feedstock. Figure 5.17(a) and (c) show SEM micrographs of A4C-B, A4C-SD and A8C-SD powders. Apparently, MWNTs are dispersed only on the surface of A4C-B powder, but are dispersed throughout the powder agglomerates of A4C-SD and A8C-SD.

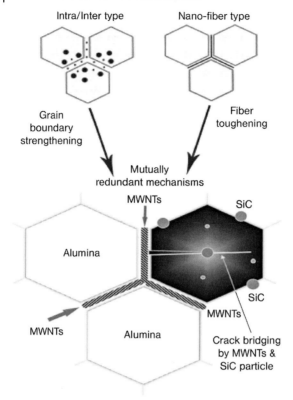

Figure 5.14 Schematic diagrams showing hybrid microstructure design of alumina reinforced with MWNTs and SiC nanoparticles. Reproduced with permission from [75]. Copyright © (2008) Elsevier.

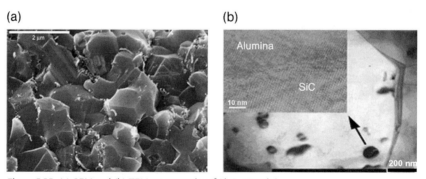

Figure 5.15 (a) SEM and (b) TEM micrographs of Al_2O_3/(5 vol% MWNT + 1 vol% SiC) hybrid. Inset in (b) is HRTEM image showing a sharp and clean SiC/alumina interface.
Reproduced with permission from [75]. Copyright © (2008) Elsevier.

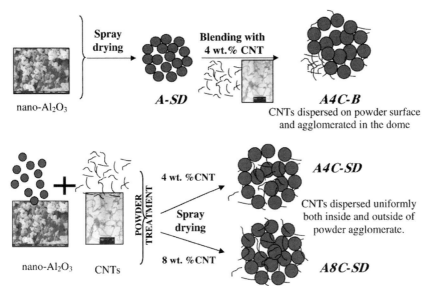

Figure 5.16 Schematic illustrations showing powder treatment for alumina nanopowders with CNT addition and dispersion. Reproduced with permission from [77]. Copyright © (2008) Elsevier.

The microstructure of plasma-sprayed alumina/MWNT nanocomposite coatings depends mainly on the powder pretreatment conditions (e.g. nanofiller content, filler dispersion) and plasma spray parameters. These in turn relate to the thermal property and kinetic velocity of plasma sprayed powders. The high thermal conductivity of MWNTs changes the heat transfer characteristic during the splat formation. The microstructure of the nanocomposite coatings consists of a fully melted zone (FM), a partially melted (PM) or solid-state sintering zone, and porosity. During plasma spraying, A4C-B spray-dried powder is superheated as a result of the high density of CNTs on the surface. Therefore, a high surface temperature of 2898 K is attained, facilitating formation of a large fraction of fully molten and re-solidified structure in the coating. For the A4C-SD spray-dried powder, CNTs are distributed uniformly both on the surface and inside powder particles (Figure 5.17(b)). This leads to reductions in both thermal exposure (2332 K) and superheating temperature ($\Delta T = -566$ K). From the temperature and velocity profiles of in-flight particles exiting from the plasma gun, Agarwal's group developed a process map for plasma spraying of alumina/MWNT nanocomposite coatings (Figure 5.18). This map provides a processing control tool for plasma spraying of nanocomposite coatings [77].

5.4.1.4 Template Synthesis

Anodic aluminum oxide (AAO) membrane having regularly arrayed nanochannels of uniform length and diameter is an ideal template to synthesize aligned CNTs via chemical vapor deposition [78–82]. AAO templates can be prepared by anodizing

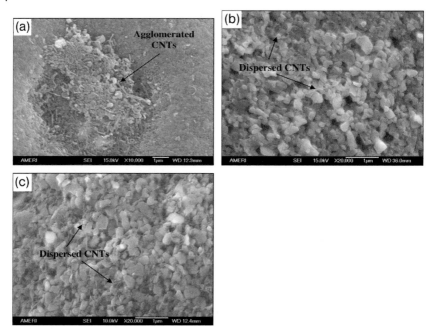

Figure 5.17 (a) Agglomeration of CNTs in A4C-B; dispersion of CNTs in (b) A4C-SD and (c) A8C-SD powder surfaces. Reproduced with permission from [77]. Copyright © (2008) Elsevier.

aluminum in various acid solutions. The density, diameter and length of pores can be controlled by varying anodizing conditions such as temperature, electrolyte, applied voltage, anodizing time, and so on. Furthermore, a multi-step anodizing process is recognized as an effective route for creating an amorphous alumina coating with widened pores (Figure 5.19(a) and (b)).

5.4.2
Silica Matrix

Silica-based ceramics are attractive materials for use in optical devices but their brittleness limits their applications. The incorporation of CNTs into silica can improve its mechanical performance markedly. Since CNTs have unique linear and nonlinear optical properties [83], silica-CNT nanocomposites show promise for photonic applications including optical switching, optical waveguides, optical limiting devices and so on [84, 85]. The nanocomposites also exhibit excellent low dielectric constant and electromagnetic shielding characteristics [86]. Silica-CNT nanocomposites can be produced by direct powder mixing [87], sol-gel [84–86, 88–92] and electrophoretic deposition (EPD) [93] processes. Among them, sol-gel is often used to make silica-CNT nanocomposites with better dispersion of CNTs in silica.

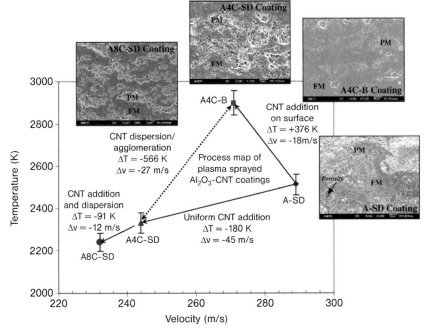

Figure 5.18 Process map of plasma sprayed coatings indicating change in temperature and velocity of in-flight particles with CNT addition and dispersion. Reproduced with permission from [77]. Copyright © (2008) Elsevier.

Sol-gel processing is a versatile technique for making high quality glasses and fine ceramics in the forms of thin films and monoliths at much lower temperatures than conventional processing techniques. The process involves the mixing of metal alkoxide precursors such as silicon tetraethyl orthosilicate (TEOS; $Si(OC_2H_5)_4$) at

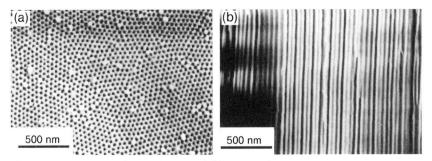

Figure 5.19 SEM images showing (a) top and (b) cross-section view of alumina template prepared by a two-step anodization method. Highly aligned nanochannels can be readily seen in (b). Reproduced with permission from [79]. Copyright © (2001) Elsevier.

the molecular level. TEOS undergoes hydrolysis and condensation leading to the formation of metal oxide network via Si−O−Si (siloxane) bonding. The presence of acid or basic catalysts can speed up the hydrolysis process. In general, CNTs must be chemically functionalized [90–92]. Otherwise, surfactants must be introduced into the sol to disperse pristine CNTs [86, 89]. Very recently, Berguiga et al. fabricated a SiO_2/CNT thin film using the sol-gel technique [92]. In the process, SiO_2 sol was prepared by mixing TEOS with hydrochloric acid and ethanol. MWNTs were purified, shortened and funtionalized with chlorine. Functionalized MWNTs were introduced into silica sol under sonication. The SiO_2/CNT thin films were then deposited on glass substrates by the dip-coating method. Figure 5.20(a) and (b) show the surface

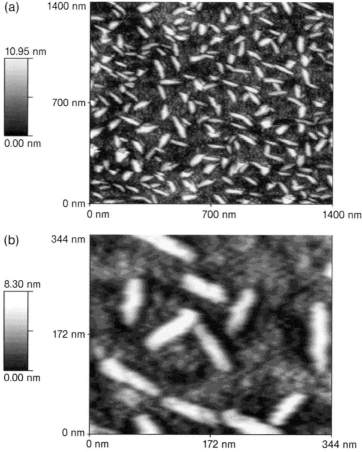

Figure 5.20 Tapping mode AFM images of silica/MWNT film prepared by the sol-gel technique with (a) square area of 1.4 μm and (b) a zoomed area. Reproduced with permission from [92]. Copyright © (2006) Elsevier.

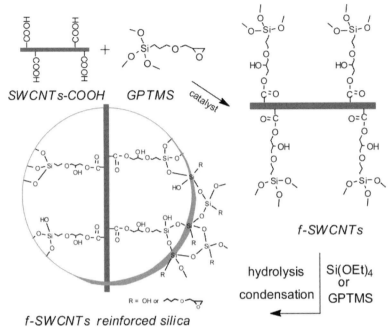

Figure 5.21 Reaction between SWNT-COOH and GPTMS in forming f-SWNT and its subsequent incorporation into silica matrix via the sol-gel technique. Reproduced with permission from [91]. Copyright © (2006) The Royal Society of Chemistry.

morphology of SiO_2/CNT thin film imaged by the atomic force microscopy (AFM) in tapping mode. Good dispersion of MWNTs in silica matrix can be readily seen in these micrographs.

Zhang et al. funtionalized SWNTs (f-SWNTs) covalently with silane agent and incorporated them into silica matrix via the sol-gel process [91]. Figure 5.21 shows the strategy of incorporating f-SWNTs into silica matrix. Pristine SWNTs were first purified with sulfuric/nitric acids to form carboxylic groups, and then reacted with (3-glycidoxypropyl)trimethoxysilane (GPTMS). The resulting mixture was stirred and heated to 70 °C for 24 h. The silica sol was prepared by hydrolysis of TEOS with ethanol, water and hydrochloric acid. GPTMS sol containing f-SWNTs was prepared by hydrolysis of the GPTMS/f-SWNT mixture with water, ethanol and formic acid. These two sols were then mixed at the same molar ratio. The hybrid sol was finally cast on a glassy carbon electrode followed by air drying. Once the functional groups are formed on the nanotube surfaces, the electrostatic repulsion between nanotubes can overcome van der Waals interactions, thereby forming a stable suspension in the solvent [91].

Loo et al. studied the effect of surfactant additions on the dispersion of MWNTs in SiO_2 sol [86]. The surfactants used include hexacetyltrimethyl ammonium bromide

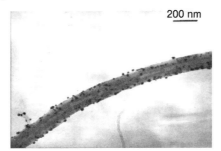

Figure 5.22 TEM micrograph showing adsorption of PAA molecules on CNTs surface. Reproduced with permission from [89]. Copyright © (2004) Elsevier.

($C_{16}TAB$), N-dimethylformaldehyde (DMF), and poly(4-vinylpyridine) (PVP-4). The PVP-4 polymer is widely recognized as an effective agent for dispersing SWNTs in alcohol or TEOS alcohol solution [94]. Ning *et al.* also used cationic $C_{16}TAB$ and anionic polyacrylic acid (PAA) surfactants to disperse MWNTs in TEOS alcohol solution [89]. Figure 5.22 shows TEM image of a MWNT with PAA adsorption. The adsorption of DMF, $C_{16}TAB$ and PVP-4 dispersant agents on nanotubes has also been reported in the literature [86]. Adsorption of surfactant such as $C_{16}TAB$ on nanotube surfaces results in steric repulsion between like cationic charges. Such repulsive forces exceed van der Waals forces of attraction, thereby improving the dispersion of nanotubes in silica sol. Figure 5.23(a) and (b) show typical SEM fractographs of the SiO_2/5 vol% MWNT composite with and without $C_{16}TAB$. Apparently, MWNTs are dispersed homogenously in the silica matrix with $C_{16}TAB$. On the other hand, nanotubes agglomerate into clusters in the composite without cationic surfactant as expected.

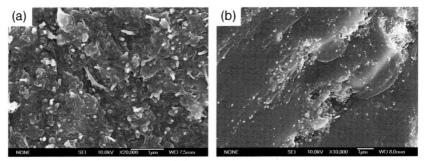

Figure 5.23 SEM fractographs of SiO_2/5 vol% MWNT composite (a) with and (b) without $C_{16}TAB$ surfactant. Reproduced with permission from [89]. Copyright © (2004) Elsevier.

5.4.3
Titania Matrix

Titania (TiO_2) is a semiconducting oxide with high photocatalytic ability. It finds application in many technological areas such as microelectronics, photocatalysis and sensors [95–97]. Titania is also an important bioceramic coating material for metal implants due to its excellent biocompatibility [98, 99]. Incorporating CNTs into titania can lead to the development of novel composite materials with advanced functional properties for photocatalytic, microelectronic and biomedical applications [100]. The techniques used for forming titania-CNT nanocomposites include heterocoagulation, sol-gel and hydrothermal treatment.

Sun and Gao employed heterocoagulation to deposit titania on MWNTs [62]. In the process, ammonia-treated MWNTs were dispersed ultrasonically in water containing cationic PEI. Titania nanoparticles with sizes less than 10 nm were dispersed in water under sonication and the pH value was adjusted by adding NaOH. Titania drops were then added to the nanotube suspension under a pH of 8. Figure 5.24 shows the zeta potential vs pH curves for titania and NH_3-treated MWNTs measured in 1 mM NaCl solution. For zeta-potential measurements, the pH was adjusted with HCl for acidic suspension and with NaOH for basic suspension. Electronegative titania nanoparticles are attracted to positively charged surfaces of MWNTs, thereby forming a titania coating layer on the nanotube surface (Figure 5.25). In another study, Sun et al. coated SWNTs with titania nanoparticles by using $TiCl_4$ precursor and PEI surfactant by the hydrothermal method [101]. Titania nanoparticles were formed by the hydrolysis of $TiCl_4$ solution through the following reaction:

$$TiCl_4 + 2H_2O \rightarrow TiO_2 + 4H^+ + 4Cl^-. \tag{5.1}$$

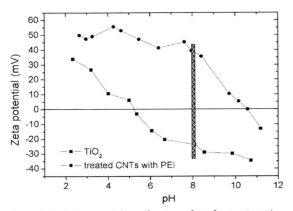

Figure 5.24 Zeta potential as a function of pH for titania and treated MWNTs with PEI. Reproduced with permission from [62]. Copyright © (2003) Elsevier.

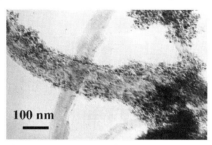

Figure 5.25 TEM image showing titania coated MWNT surface. Reproduced with permission from [62]. Copyright © (2003) Elsevier.

Acid-purified SWNTs were then added to the $TiCl_4$ suspension. PEI addition is beneficial to assist the deposition of titania on nanotube surfaces.

Jitianu et al. employed both sol-gel and hydrothermal techniques to coat titania on MWNTs [102]. For the sol-gel process, alkoxide precursor, that is, titanium tetra-ethoxide $Ti(OEt)_4$ or titanium tetra-isopropoxide $Ti(OPr)_4$ was mixed with ethanol/isopropanol, water and nitric acid to form a titania sol under magnetic stirring. MWNTs were then mixed with the sol under stirring. The impregnated MWNTs were separated from the solution by filtration. For hydrothermal treatment, the precursor consisted of 15 wt% $TiOSO_4$ in diluted sulfuric acid. MWNTs were then added to the $TiOSO_4$ solution in water. Hydrothermal treatment was performed in an autoclave at 120 °C for 5 h. Impregnated nanotubes prepared by the sol-gel and hydrothermal methods were further dried at high temperatures. MWNT surfaces prepared by sol-gel using $Ti(OEt)_4$ precursor are coated with a continuous thin film (Figure 5.26(a)), and with TiO_2 nanoparticles in the case of $Ti(OPr)_4$ (Figure 5.26(b)). The microstructure of titania coated nanotubes prepared by hydrothermal method is shown in Figure 5.26(c).

5.4.4
Zirconia Matrix

Zirconia-based ceramics are well recognized for their excellent mechanical, electrical, thermal and optical properties. They find a broad range of industrial applications including oxygen sensors, solid oxide fuel cells and ceramic membranes [103, 104]. The physical and mechanical properties of zirconia can be further enhanced by adding low loading level of CNTs. Zirconia-CNT nanocomposites can be fabricated by hot-pressing [105], heterocoagulation [106] and hydrothermal crystallization [107, 108].

Shan and Gao fabricated ZrO_2/MWNT nanocomposites by hydrothermal treatment of $ZrOCl_2 \cdot 8H_2O$. They reported that full coverage of ZrO_2 on nanotubes can be prepared only through the use of an acid medium [108]. In a weak alkaline solution, only few zirconia agglomerates adhere to the sidewalls of MWNTs. Hardly any adsorption of zirconia on nanotubes can be observed in a strong alkaline solution. They attributed the formation of zirconia coating on sidewalls of MWNTs to the

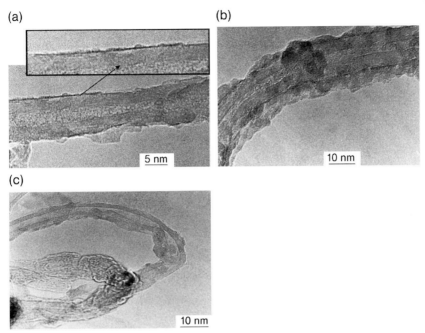

Figure 5.26 TEM images of titania coated MWNT composite prepared by sol-gel method using (a) Ti(OEt)$_4$ and (b) Ti(OPr)$_4$ precursor. Inset in (a) is higher magnification view of coating layer. (c) MWNTs coated with TiO$_2$ prepared by hydrothermal method. Reproduced with permission from [102]. Copyright © (2004) Elsevier.

electrostatic attraction between positively charged ZrOCl$_2$.8H$_2$O precursor and negatively charged MWNTs.

5.5
Carbide-Based Nanocomposites

5.5.1
Silicon Carbide Matrix

Non-oxide ceramics such as silicon carbide, boron carbide and silicon nitride are attractive structural materials for high temperature applications because of their low density, very high hardness, excellent thermal and chemical stability [109–111]. Despite these advantages they are susceptible to fast fracture during mechanical loading due to their inherently brittle nature. The incorporation of metal nanoparticles into such ceramics can mitigate the problems associated with brittleness. Carbon nanotubes with high aspect ratios are excellent reinforcing and toughening materials for improving the toughness of non-oxide ceramics.

Silicon carbide exhibits different crystalline structures from hexagonal (α-SiC), cubic (β-SiC) to rhombohedral. Of these, cubic β-SiC is particularly important because of its higher bending strength, hardness, stiffness and fracture toughness when compared with α-SiC. Because of the high melting point of silicon carbide, PM method becomes the primary processing technique for making ceramic products. Further, silicon carbide exhibits poor sinterability due to its strong covalent bonding and high melting point. Thus, sintering aids must be added to obtain dense ceramic specimens. SiC-CNT nanocomposites have been fabricated by spray pyrolysis [112], conventional powder mixing followed by hot pressing [41] or by SPS [113, 114], microwave synthesis [115] and preceramic polymer precursor methods [116]. Spray pyrolysis of xylene suspension containing ferrocene and SiC powders into a reactor at 1000 °C produces SiC composite flakes with uneven distribution of nanotubes [112]. Ma et al. [41] fabricated the (SiC + 1% B_4C)/10%CNT nanocomposite by blending SiC nanopowder (80 nm), CNT and sintering aid (B_4C) in butylalcohol ultrasonically, followed by hot-pressing at 2000 °C for 1 h. High temperature is required for good consolidation of these powders into bulk nanocomposite. However, high temperature sintering can result in severe grain growth and destruction of the integrity of nanotubes.

Spark plasma sintering with very short processing time is an alternative consolidation route for SiC-CNT nanocomposites. However, SPS of SiC-based materials must be carried out at temperatures $\geq$1800 °C due to the strong covalent bonding of ceramics. Hirota et al. studied the effect of SPS temperature on the microstructure of monolithic SiC and its composites reinforced with carbon nanofibers [114]. Monolithic SiC and its composites were prepared by direct mixing of powder constituents in methyl alcohol followed by ball milling and SPS. Figure 5.27 shows the density and average grain size of monolithic β-SiC as a function of plasma sintering temperature. The relative density of β-SiC sintered at 1700 °C is only 80.9% but increases to 96.4% at 1800 °C. The density saturates with further increasing temperature. Further, the grain size of β-SiC increases with increasing temperature as expected. The incorporation of carbon nanofibers into monolithic SiC has little effect on the relative density with the value kept at about 96%. However, the grain size decreases sharply from ~4.2 μm to 1.3 μm by adding 5 vol% VGCF, and reduces slightly to ~1.2 μm at 15 vol% VGCF (Figure 5.28).

To preserve the properties of CNTs, it is advantageous to fabricate the SiC/CNT nanocomposites at lower temperatures, about 1300 °C. In this regard, polymer derived ceramics (PDCs) show promise to make such composites at relatively lower temperatures. This process involves the initial cross-linking of polymer precursors followed by a thermal induced polymer to ceramic transformation. The networks of organoelement compounds are directly converted into covalent bonded ceramics during pyrolysis. Typical polymer precursors commonly used include polycarbosilane (PCS), polysiloxane (PSO), and polysilazane (PSZ), producing amorphous Si–C, Si–O–Si and Si–C–N ceramics respectively upon pyrolysis. PDCs offer several advantages over conventional powder mixing and sintering in terms of the ease of controlling the structure of ceramics by designing the chemistry of polymer precursors, homogeneous chemical distribution at molecular level, near net shape

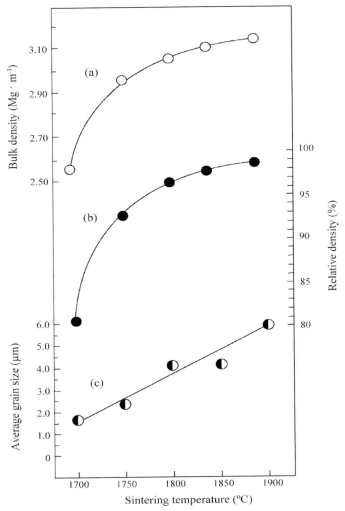

Figure 5.27 (a) Bulk density, (b) relative density and (c) average grain size vs sintering temperature for monolithic SiC prepared by SPS of Si, C, and B powder mixtures. Reproduced with permission from [114]. Copyright © (2007) Elsevier.

forming and low processing cost [56]. Major drawbacks of PDCs are large weight loss and shrinkage of ceramics during pyrolysis. During the nanocomposite fabrication, CNTs are dispersed in liquid-phase polymer under sonication, followed by pyrolysis of dried powder mixture. Yamamoto et al. [116] prepared SiC/SWNT nanocomposites by pyrolysis PCS precursor at 1400 °C in vacuum. SiC nanoparticles of 14.6–22.4 nm were deposited onto SWNTs. An et al. [56] and Katsuda et al. [117] fabricated Si-C-N/MWNT nanocomposites by using PSZ precursor. PSZ possesses amorphous network which transforms to amorphous silicon carbonitride ceramics upon pyrolysis at 1000–1100 °C [118].

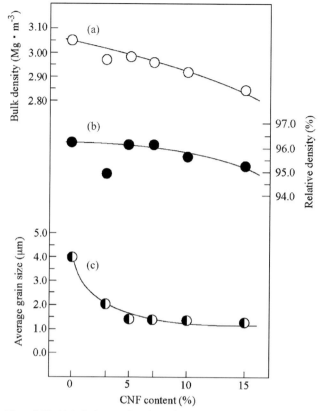

Figure 5.28 (a) Bulk density; (b) relative density; (c) average grain size vs carbon nanofiber content, for SiC/VGCF nanocomposites prepared by SPS of Si, C, B and VGCF powder mixtures at 1800 °C. Reproduced with permission from [114]. Copyright © (2007) Elsevier.

5.6
Nitride-Based Nanocomposites

5.6.1
Silicon Nitride Matrix

It is recognized that silicon nitride is difficult to sinter and consolidate into a dense material using conventional sintering processes. The incorporation of CNTs into silicon nitride would further impair its sinterability. Balazsi et al. fabricated the Si_3N_4/1% MWNT nanocomposite using hot isostatic pressing and SPS [43, 119, 120]. For the hipping process, ball-milled composite powder mixtures were sintered at 1700 °C under a pressure of 20 MPa for 3 h, and under 2 MPa for 1 h, respectively. In the case of SPS, ball-milled composite powder mixtures were consolidated at 1500 °C under 100 MPa for 3 min, and at 1650 °C under 50 MPa for 3 min, respectively.

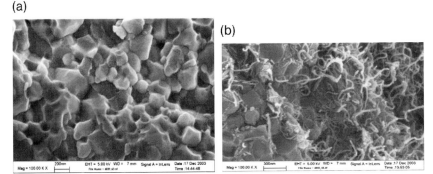

Figure 5.29 SEM fractographs of spark plasma sintered (a) Si_3N_4 at 1500 °C for 3 min under 50 MPa and (b) Si_3N_4/1 wt% MWNT nanocomposite at 1500 °C for 3 min under 100 MPa. Reproduced with permission from [120]. Copyright © (2005) Elsevier.

They reported that CNTs were missing in the nanocomposite hipped at 20 MPa for 3 h due to the destruction of CNTs during prolonged processing treatment. Carbon nanotubes survived in the nanocomposite hipped at 2 MPa for 1 h. However, the composite is porous having coarser grain structure [120]. On the other hand, SPS results in denser composites with nanocrystalline structure (Figure 5.29(a) and (b)). Ball milling MWNT, Si_3N_4 and sintering aids in ethanol under sonication do not provide sufficient dispersion of CNTs. MWNTs are mainly located in intergranular regions, and clustering of nanotubes can be readily seen in the micrographs.

References

1 Zhang, S., Sun, D., Fu, Y. and Du, H. (2005) Toughening of hard nanostructured thin films: A critical review. *Surface & Coatings Technology*, **198**, 2–8.

2 Gupta, T.K., Bechtold, J.H., Kuznicki, R.C., Cadoff, L.H. and Rossing, B.R. (1977) Stabilization of tetragonal phase in polycrystalline zirconia. *Journal of Materials Science*, **12**, 2421–2426.

3 Heuer, A.H., Claussen, N., Kriven, W.M. and Ruhle, M. (1982) Stability of tetragonal ZrO_2 particles in ceramic matrices. *Journal of the American Ceramic Society*, **65**, 642–650.

4 Heuer, A.H. (1987) Transformation toughening in ZrO_2-containing ceramics. *Journal of the American Ceramic Society*, **70**, 689–698.

5 Budiansky, B., Amazigo, J.C. and Evans, A.G. (1988) Small-scale crack bridging and the fracture toughness of particulate-reinforced ceramics. *Journal of the Mechanics and Physics of Solids*, **36**, 167–187.

6 Mataga, P.A. (1989) Deformation of crack-bridging ductile reinforcements in toughened brittle materials. *Acta Metallurgica*, **37**, 3349–3359.

7 Bao, G. and Hui, C.Y. (1990) Effects of interface debonding on the toughness of

ductile-particle reinforced ceramics. *International Journal of Solids and Structures*, **26**, 631–642.
8 Agrawal, P. and Sun, C.T. (2004) Fracture in metal-ceramic composites. *Composites Science and Technology*, **64**, 1167–1178.
9 Zimmermann, A., Hoffman, M., Emmel, T., Gross, D. and Rodel, J. (2001) Failure of metal-ceramic composites with spherical inclusions. *Acta Materialia*, **49**, 3177–3187.
10 Knowles, K.M. and Turan, S. (2002) Boron nitride-silicon carbide interphase boundaries in silicon nitride-silicon carbide particulate composites. *Journal of the European Ceramic Society*, **22**, 1587–1600.
11 Liu, Y.S., Cheng, L., Zhang, L., Hua, Y. and Yang, W. (2008) Microstructure and properties of particle reinforced silicon and silicon nitride ceramic matrix composites prepared by chemical vapor infiltration. *Materials Science and Engineering A*, **475**, 217–223.
12 Hua, Y., Zhang, L., Cheng, L. and Wang, J. (2006) Silicon carbide whisker reinforced silicon carbide composites by chemical vapor infiltration. *Materials Science and Engineering A*, **428**, 346–350.
13 Park, K. and Vasilos, T. (1998) Interface and thermal shock resistance of SiC fiber/SiC composites. *Scripta Materialia*, **39**, 1593–1598.
14 Zhu, S., Mizuno, M., Kagawa, Y. and Mutoh, Y. (1999) Monotonic tension, fatigue and creep behavior of SiC-fiber-reinforced SiC-matrix composites: A review. *Materials Science and Engineering A*, **59**, 833–851.
15 Dong, S.M., Katoh, Y., Kohyama, A., Schwab, S.T. and Snead, L.L. (2002) Mirostructural evolution and mechanical performances of SiC/SiC composites by polymer impregnation/microwave pyrolysis (PIMP) process. *Ceramics International*, **28**, 899–905.
16 Li, B., Zhang, C.R., Cao, F., Wang, S.Q., Cao, C.B., Feng, J. and Chen, B. (2008) Fabrication of high density three-dimensional carbon fiber reinforced nitride composites by precursor infiltration and pyrolysis. *Advances in Applied Ceramics*, **107**, 1–3.
17 Rocha, R.M., Cairo, C.A. and Graca, M.L. (2006) Formation of carbon-fiber-reinforced ceramics matrix composites with polysiloxane/silicon derived matrix. *Materials Science and Engineering A*, **437**, 268–273.
18 Mishra, R..S., Lesher, C.E. and Mukherjee, A.K. (1996) High-pressure sintering of nanocrystalline γ-Al_2O_3. *Journal of the American Ceramic Society*, **79**, 2989–2992.
19 Zhan, G.D., Kuntz, J., Wan, J., Garay, J. and Mukherjee, A.K. (2002) Alumina-based nanocomposites consolidated by spark plasma sintering. *Scripta Materialia*, **47**, 737–741.
20 Kim, B.N., Hiraga, K., Morita, K. and Yoshida, H. (2007) Spark plasma sintering of transparent alumina. *Scripta Materialia*, **57**, 607–610.
21 Hiraga, K., Kim, B.N., Morita, K., Yoshida, H., Suzuki, T.S. and Sakka, Y. (2007) High-strain-rate superplasticity in oxide ceramics. *Science and Technology of Advanced Materials*, **8**, 578–587.
22 Zhou, X., Hulbert, D.L., Kuntz, J.D., Sadangi, R.K., Shukla, V., Kear, B.H. and Mukherjee, A.M. (2005) Superplasticity of zirconia-alumina-spinel nanoceramic composite by spark plasma sintering of plasma sprayed powders. *Materials Science and Engineering A*, **394**, 353–359.
23 Niihara, K. (1991) New design concept of structural ceramics-ceramic nanocomposites. *Journal of the Ceramic Society of Japan*, **99**, 974–982.
24 Niihara, K. and Nakahira, A. (1992) Sintering behaviors and consolidation process for Al_2O_3/SiC nanocomposites. *Journal of the Ceramic Society of Japan*, **100**, 448–453.
25 Anya, C.C. (1999) Microstructural nature of strengthening and toughening in

26 Gao, L., Wang, H.Z., Hong, J.S., Miyamoto, H., Miyamoto, K., Nishikawa, Y. and Torre, S.D. (1999) Mechanical properties and microstructures of nano-SiC-Al$_2$O$_3$ composites densified by spark plasma sintering. *Journal of the European Ceramic Society*, **19**, 609–613.

27 Galusek, D., Sedlacek, J., Svancarek, P., Riedel, R., Satet, R. and Hoffman, M. (2007) the influence of post-sintering HIP on the microstructure, hardness and indentation fracture toughness of polymer-derived Al$_2$O$_3$-SiC nanocomposites. *Journal of the European Ceramic Society*, **27**, 1237–1245.

28 Tan, H. and Yang, W. (1998) Toughening mechanisms of nanocomposite ceramics. *Mechanics of Materials*, **30**, 111–123.

29 Zhang, Y., Wang, L., Jiang, W., Chen, L. and Bai, G. (2006) Microstructure and properties of Al$_2$O$_3$-TiC nanocomposites fabricated by spark plasma sintering from high-energy ball milled reactants. *Journal of the European Ceramic Society*, **26**, 3393–3397.

30 Zhu, W.Z., Gao, J.H. and Ding, Z.S. (1997) Microstructure and mechanical properties of a Si$_3$N$_4$/Al$_2$O$_3$ nanocomposite. *Journal of Materials Science*, **32**, 537–542.

31 Balog, M., Keckes, J., Schoberl, T., Galusek, D., Hofer, F., Krest'an, J., Lences, Z., Huang, J.L. and Sajgalik, P. (2007) Nano/macro-hardness and fracture resistance of Si$_3$N$_4$/SiC nanocomposites with up to 13 wt% of SiC nanoparticles. *Journal of the European Ceramic Society*, **27**, 2145–2152.

32 Isobe, T., Daimon, K., Ito, K., Matsubara, T., Hikichi, Y. and Ota, T. (2007) Preparation and properties of Al$_2$O$_3$/Ni composite from NiAl$_2$O$_4$ spinel by in-situ reaction sintering method. *Ceramics International*, **33**, 1211–1215.

33 Isobe, T., Daimon, K., Sato, T., Matsubara, T., Hikichi, Y. and Ota, T. (2008) Spark plasma sintering technique for reaction sintering of Al$_2$O$_3$/Ni nanocomposite and its mechanical properties. *Ceramics International*, **34**, 213–217.

34 Bhaduri, S. and Bhaduri, S.S. (1997) Enhanced low temperature toughness of Al$_2$O$_3$- ZrO$_2$ nano/nano composites. *Nanostruct Mater*, **8**, 755–763.

35 Mishra, R.S. and Mukherjee, A.K. (2001) Processing of high hardness-high toughness alumina matrix nanocomposites. *Materials Science and Engineering A*, **301**, 97–101.

36 Qiu, L.X., Yao, B., Ding, Z.H., Zheng, Y.Z., Jia, X.P. and Zheng, W.T. (2008) Characterization of structure and properties of TiN-TiB$_2$ nanocomposite prepared by ball milling and high pressure heat treatment. *Journal of Alloys and Compounds*, **456**, 436–440.

37 Sternitzke, M. (1997) Review: structural ceramic nanocomposites. *Journal of the European Ceramic Society*, **17**, 1061–1082.

38 Huang, J.Y., Chen, S., Wang, Z.Q., Kempa, K., Wang, Y.M., Jo, S.H., Chen, G., Dresselhaus, M.S. and Ren, Z.F. (2006) Superplastic carbon nanotubes. *Nature*, **439**, 281.

39 Peigney, A., Flahaut, F., Laurent, C., Chastel, F. and Rousset, A. (2002) Aligned carbon nanotubes in ceramic-matrix nanocomposites prepared by high-temperature extrusion. *Chemical Physics Letters*, **352**, 20–25.

40 Zhan, G., Kuntz, J.D. and Mukherjee, A.K. (2005) Anisotropic thermal applications of composites of ceramics and carbon nanotubes. US Patent 6976532.

41 Ma, R.Z., Wu, J., Wei, B.Q., Liang, J. and Wu, D.H. (1998) Processing and properties of carbon nanotubes-nano-SiC ceramic. *Journal of Materials Science*, **33**, 5243–5246.

42 Laurent, C., Peigney, A., Dumortier, O. and Rousser, A. (1998) Carbon nanotubes-Fe-alumina nanocomposites. Part II: Microstructure and mechanical properties of the hot pressed composites. *Journal of the European Ceramic Society*, **18**, 2005–2013.

43 Balazsi, C., Weber, F., Kover, Z., Chen, Z., Konya, Z., Kasztovsky, Z., Vertesy, Z., Biro, L.P., Kiricsi, I. and Arato, P. (2006) Application of carbon nanotubes to silicon nitride matrix reinforcements. *Current Applied Physics*, **6**, 124–130.

44 Zhan, G.D., Kuntz, J.D., Wan, J. and Mukherjee, A.K. (2003) Single-wall carbon nanotubes as attractive toughening agents in alumina-based nanocomposites. *Nature Materials*, **2**, 38–42.

45 Li, J.L., Bai, G.Z., Feng, J.W. and Jiang, W. (2006) Microstructure and mechanical properties of hot-pressed carbon nanotubes compacted by spark plasma sintering. *Carbon*, **43**, 2649–2653.

46 Liu, J.Q., Xiao, T., Liao, K. and Wu, P. (2007) Interfacial design of carbon nanotube polymer composites: A hybrid system of noncovalent and covalent functionalizations. *Nanotechnology*, **18**, 1657011–1657016.

47 Hernadi, K., Couteau, E., Seo, J.W. and Forro, L. (2003) $Al(OH)_3$/multiwalled carbon nnaotube composite: Homogeneous coverage of $Al(OH)_3$ on carbon nanotube surfaces. *Langmuir*, **19**, 7026–7029.

48 Morisada, Y. and Miyamoto, Y. (2004) SiC-coated carbon nanotubes and their application as reinforcements for cemented carbides. *Materials Science and Engineering A*, **381**, 57–61.

49 Guo, Y., Cho, H., Shi, D., Lian, J., Song, Y., Abot, J., Poudel, B., Ren, Z., Wang, L. and Ewing, R.C. (2007) Effects of plasma surface modification on interfacial behaviors and mechanical properties of carbon nanotube-Al_2O_3 nanocomposites. *Applied Physics Letters*, **91**, 2619031–2619033.

50 Hwang, G.L. and Hwang, K.C. (2001) Carbon nanotube reinforced ceramics. *Journal of Materials Chemistry*, **11**, 1722–1725.

51 Cha, S.I., Kim, K.T., Lee, K.H., Mo, C.B. and Hong, S.H. (2005) Strengthening and toughening of carbon nanotube reinforced alumina nanocomposite fabricated by molecular level mixing process. *Scripta Materialia*, **53**, 793–797.

52 Sun, J., Gao, L. and Jin, X. (2005) Reinforcement of alumina matrix with multiwalled carbon nanotubes. *Ceramics International*, **31**, 893–896.

53 Poyato, R., Vasiliev, A.L., Padture, N.P., Tanaka, H. and Nishimura, T. (2006) Aqueous colloidal processing of single-wall carbon nanotubes and their composites with ceramics. *Nanotechnology*, **17**, 1770–1777.

54 Fan, J.P., Zhuang, D.M., Zhao, D.Q., Zhang, G. and Wu, M.S. (2006) Toughening and reinforcing alumina matrix composite with single-wall carbon nanotubes. *Applied Physics Letters*, **89**, 1219101–1219103.

55 Fan, J.P., Zhao, D.Q., Wu, M.S., Xu, Z. and Song, J. (2006) Preparation and microstructure of multiwalled carbon nanotubes-toughened composite. *Journal of the American Ceramic Society*, **89**, 750–753.

56 Bharadwaj, L., Fan, Y., Zhang, L., Jiang, D. and An, L. (2004) Oxidation behavior of a fully dense polymer-derived amorphous silicon carbonitride ceramic. *Journal of the American Ceramic Society*, **87**, 483–486.

57 Siegel, R.W., Chang, S.K., Ash, B.J., Stone, J., Ajayan, P.M., Doremus, R.W. and Schadler, L.S. (2001) Mechanical behavior of polymer and ceramic matrix nanocomposites. *Scripta Materialia*, **44**, 2061–2064.

58 Peigney, A., Laurent, C., Dumortier, O. and Rousser, A. (1998) Carbon nanotubes-Fe-alumina nanocomposites. Part I: Influence of the Fe content on the synthesis of powders. *Journal of the European Ceramic Society*, **18**, 1995–2004.

59 Flahaut, E., Peigney, A., Laurent, C., Marliere, C., Chastel, F. and Rousset, A. (2000) Carbon nanotube-metal-oxide nanocomposites: Microstructure, electrical conductivity and mechanical

properties. *Acta Materialia*, **48**, 3803–3812.

60 Peigney, A., Laurent, C., Flahaut, E. and Rousset, A. (2000) Carbon nanotubes in novel ceramic matrix composites. *Ceramics International*, **26**, 677–683.

61 Sun, J., Gao, L. and Li, W. (2002) Colloidal processing of carbon nanotube/alumina composites. *Chemistry of Materials*, **14**, 5169–5172.

62 Sun, J. and Gao, L. (2003) Development of a dispersion process for carbon nanotubes in ceramic matrix by heterocoagulation. *Carbon*, **41**, 1063–1068.

63 Lu, K. (2008) Freeze cast carbon nanotube-alumina nanoparticles green composites. *Journal of Materials Science*, **43**, 652–659.

64 Lewis, A. (2000) Colloidal processing of ceramics. *Journal of the American Ceramic Society*, **83**, 2342–2359.

65 Cesarano, J., Aksay, I.A. and Bleier, A. (1988) Stability of aqueous alpha-alumina suspensions with poly(methacrylic acid) polyelectrolyte. *Journal of the American Ceramic Society*, **71**, 250–255.

66 http://www.malvern.com/LabEng/technology/zeta_potential/zeta_potential_LDE.htm.

67 Jiang, L., Gao, L. and Sun, J. (2003) Production of aqueous colloidal dispersions of carbon nanotubes. *Journal of Colloid and Interface Science*, **260**, 89–94.

68 Wei, T., Fan, Z., Luo, G. and Wei, F. (2008) A new structure for multi-walled carbon nanotubes reinforced alumina nanocomposite with high strength and toughness. *Materials Letters*, **62**, 641–644.

69 Thomson, K., Jiang, D., Ritchie, R.O. and Mukherjee, A.K. (2007) A preservation of carbon nanotubes in alumina-based nanocomposites via Raman spectroscopy and nuclear magnetic resonance. *Applied Physics A-Materials Science & Processing*, **89**, 651–654.

70 Jiang, D., Thomson, K., Kuntz, J.D., Ager, J.W. and Mukherjee, A.K. (2007) Effect of sintering temperature on a single-wall carbon nanotube-toughened alumina-based nanocomposite. *Scripta Materialia*, **56**, 959–962.

71 Yamamoto, G., Omori, M., Hashida, T. and Kimura, H. (2008) A novel structure for carbon nanotube reinforced alumina composites with improved mechanical properties. *Nanotechnology*, **19**, 3157081–3157087.

72 Cha, S.I., Kim, K.T., Lee, K.H. Mo, C.B. and Hong, S.H. (2005) Strengthening and toughening of carbon nanotube reinforced alumina nanocomposite fabricated by molecular level mixing process. *Scripta Materialia*, **53**, 793–797.

73 Vasiliev, A.L., Poyato, R. and Padture, N.P. (2007) Single-wall carbon nanotubes at grain boundaries. *Scripta Materialia*, **56**, 461–463.

74 Estili, M. and Kawasaki, A. (2008) An approach to mass-producing individually alumina-decorated multiwalled carbon nanotubes with optimized and controlled compositions. *Scripta Materialia*, **58**, 906–909.

75 Ahmad, K. and Pan, W. (2008) Hybrid nanocomposites: A new route towards tougher alumina ceramics. *Composites Science and Technology*, **68**, 1321–1327.

76 Balani, K., Bakshi, S.R., Chen, Y., Laha, T. and Agarwal, A. (2007) Role of powder treatment and carbon nanotube dispersion in the fracture toughening of plasma-sprayed aluminum oxide-carbon nanotube nanocomposite. *Journal of Nanoscience and Nanotechnology*, **7**, 3553–3562.

77 Balani, K. and Agarwal, A. (2008) Process map for plasma sprayed aluminum oxide-carbon nanotube nanocomposite coatings. *Surface & Coatings Technology*, **202**, 4270–4277.

78 Sui, Y.C., Acosta, D.R., Gonzalez-Leon, J.A., Bermudez, A., Feuchtwanger, J., Cui, B.Z., Flores, J.O. and Saniger, J.M. (2001) Structure, thermal stability, and deformation of multibranched carbon nanotubes synthesized by CVD in the

AAO template. *The Journal of Physical Chemistry. B*, **105**, 1523–1527.

79 Zhang, X.Y., Zhang, L.D., Zheng, M.J., Li, G.H. and Zhao, L.X. (2001) Template synthesis of high-density carbon nanotubes arrays. *Journal of Crystal Growth*, **223**, 306–310.

80 Im, Y.S., Cho, Y.S., Choi, G.S., Yu, F.C. and Kim, D.J. (2004) Stepped carbon nnaotube synthesized in anodic aluminum oxide templates. *Diamond and Related Materials*, **13**, 1214–1217.

81 Xia, Z., Riester, L., Curtin, W.A., Li, H., Sheldon, B.W., Liang, J., Chang, B. and Xu, J.M. (2004) Direct observation of toughening mechanisms in carbon nanotube ceramic matrix composites. *Acta Materialia*, **52**, 931–944.

82 Schneider, J.J., Maksimova, N.I., Engstler, J., Joshi, R., Schierholz, R. and Feile, R. (2008) Catalyst free growth of a carbon nanotube-alumina composite structure. *Inorganica Chimica Acta*, **361**, 1770–1778.

83 Guo, G.Y., Chu, K.C., Wang, D.S. and Duan, C.G. (2004) Linear and nonlinear optical properties of carbon nanotubes from first principle calculations. *Physical Review B-Condensed Matter*, **69**, 205416(1)–205416(11).

84 DiMaio, J., Rhyne, S., Yang, Z., Fu, K., Czerw, R., Xu, J., Webster, S., Sun, Y.P., Carroll, D.L. and Ballato, J. (2003) Transparent silica glasses containing single walled carbon nanotubes. *Information Sciences*, **149**, 69–73.

85 Zhan, H.B., Zheng, C., Chen, W. and Wang, M. (2005) Characterization and nonlinear optical property of a multi-walled carbon nanotube/silica xerogel composite. *Chemical Physics Letters*, **411**, 373–377.

86 Loo, S., Idapalapati, S., Wang, S., Shen, L. and Mhaisalkar, S.G. (2007) Effect of surfactants on MWCNT-reinforced sol-gel silica dielectric composites. *Scripta Materialia*, **57**, 1157–1160.

87 Ning, J., Zhang, J., Pan, Y. and Guo, J. (2003) Fabrication and mechanical properties of SiO_2 matrix composites reinforced by carbon nanotube. *Materials Science and Engineering A*, **357**, 392–396.

88 Seeger, T., Kohler, T., Frauenheim, T., Grobert, N., Ruhle, M., Terrones, M. and Seifert, G. (2002) Nanotube composites: Novel SiO_2 coated carbon nanotubes. *Chemical Communications*, **1**, 34–35.

89 Ning, J., Zhang, J., Pan, Y. and Guo, J. (2004) Surfactants assisted processing of carbon nanotube-reinforced SiO_2 matrix composites. *Ceramics International*, **30**, 63–67.

90 de Andrade, M.J., Lima, M.D., Bergmann, C.P., Ramminger, G., Balzaretti, N.M., Costa, T.M. and Gallas, M.R. (2008) Carbon nanotube/silica composites obtained by sol-gel and high pressure techniques. *Nanotechnology*, **19**, 265607(1)–265607(7).

91 Zhang, Y., Shen, Y., Han, D., Wang, Z., Song, J. and Niu, L. (2006) Reinforcement of silica with single-walled carbon nanotubes through covalent functionalization. *Journal of Materials Chemistry*, **16**, 4592–4597.

92 Berguiga, L., Bellessa, J., Vocanson, F., Bernstein, E. and Planet, J.C. (2006) Carbon nanotube silica glass composite in thin films by the sol-gel technique. *Optical Materials*, **28**, 167–171.

93 Chicatun, F., Cho, J., Schaab, S., Brusatin, G., Colombo, P., Roether, J.A. and Boccaccini, A.R. (2007) Carbon nanotube deposits and CNT/SiO_2 composite coatings by electrophoretic deposition. *Advances in Applied Ceramics*, **106**, 186–195.

94 Rouse, J.H. (2005) Polymer-assisted dispersion of single-walled carbon nanotubes in alcohols and applicability toward carbon nanotube/sol-gel composite formation. *Langmuir*, **21**, 1055–1061.

95 Francioso, L., Prato, M. and Siciliano, P. (2008) Low-cost electronics and thin film technology for sol-gel titania lambda probes. *Sensors and Actuators B-Chemical*, **128**, 359–365.

96 Zhao, L., Yu, Y., Song, L., Hu, X. and Larbot, A. (2005) Synthesis and characterization of nanostructured titania film for photocatalysis. *Applied Surface Science*, **239**, 285–291.

97 Sotter, E., Vilanova, X., Llobet, E., Vasieliev, A. and Correig, X. (2007) thick film titania sensors for detecting traces of oxygen. *Sensors and Actuators B-Chemical*, **127**, 567–579.

98 Milella, E., Cosentino, F., Licciulli, A. and Massaro, C. (2001) Preparation and characterization of titania/hydroxyapatite composite coatings by sol-gel process. *Biomaterials*, **22**, 1425–1431.

99 Wen, C.E., Xu, W., Hu, W.Y. and Hodgson, P.D. (2007) Hydroxyapatite/titania sol-gel coatings on titanium-zirconium alloy for biomedical applications. *Acta Biomaterialia*, **3**, 403–410.

100 Xia, X.H., Jia, Z.J., Yu, Y., Liang, Y., Wang, Z. and Ma, L.L. (2007) Preparation of multi-walled nanotubes supported TiO_2 and its photocatalytic activity in the reduction of CO_2 with H_2O. *Carbon*, **45**, 717–721.

101 Sun, J., Iwasa, M., Gao, L. and Zhang, Q. (2004) Single-walled carbon nanotubes coated with titania nanoparticles. *Carbon*, **42**, 895–899.

102 Jitianu, A., Cacciaguerra, T., bBenoit, R., Delpeux, S., Bbeguin, F. and Bonnamy, S. (2004) Synthesis and characterization of carbon nanotubes-TiO_2 nanocomposites. *Carbon*, **42**, 1147–1151.

103 Li, Y., Xie, Y., Gong, J., Chen, Y. and Zhang, Z. (2001) Preparation of Ni/YSZ materials for SOFC anodes by buffer-solution method. *Materials Science and Engineering: B*, **86**, 119–122.

104 Lenormand, P., Caravaca, D., Laberty-Robert, C. and Ansart, F. (2005) Thick films of YSZ electrolyte by dip-coating process. *Journal of the European Ceramic Society*, **25**, 2643–2646.

105 Duszova, A., Dusza, J., Tomasek, K., Blugan, G. and Kuebler, J. (2008) Microstructure and properties of carbon nanotube/zirconia composite. *Journal of the European Ceramic Society*, **28**, 1023–1027.

106 Sun, J., Gao, L., Iwasa, M., Nakayama, T. and Niihara, K. (2005) Failure investigation of carbon nanotube/3Y-TZP nanocomposites. *Ceramics International*, **31**, 1131–1134.

107 Lupo, F., Kamalakaran, R., Scheu, C., Grobert, N. and Ruhle, M. (2004) Microstructural investigations on zirconium oxide-carbon nanotube composites synthesized by hydrothermal crystallization. *Carbon*, **42**, 1995–1999.

108 Shan, Y. and Gao, L. (2005) Synthesis and characterization of phase controllable ZrO_2-carbon nanotubes nanocomposites. *Nanotechnology*, **16**, 625–630.

109 Jin, H.Y., Ishiyama, M., Qiao, G.J., Gao, J.Q. and Jin, Z.H. (2008) Plasma active sintering of silicon carbide. *Materials Science and Engineering A*, **484**, 270–272.

110 Chen, Y.X., Li, J.T. and Du, J.S. (2008) Cost effective combustion synthesis of silicon nitride. *Materials Research Bulletin*, **43**, 1598–1606.

111 Heian, E.M., Khalsa, S.K., Lee, W., Munir, Z.A., Yamamoto, T. and Ohyanagi, M. (2004) Synthesis of dense, high-defect concentration B_4C through mechanical activation and filed-assisted combustion. *Journal of the American Ceramic Society*, **87**, 779–783.

112 Kamalakaran, R., Lupo, F., Grobert, N., Scheu, T., Jin-Phillipp, N.Y. and Ruhle, M. (2004) Microstructural characterization of C-SiC-carbon nanotube composite flake. *Carbon*, **42**, 1–4.

113 Morisada, Y., Miyamoto, Y., Takaura, Y., Hirota, K. and Tamari, N. (2007) Mechanical properties of SiC composites incorporating SiC-coated multi-walled carbon nanotubes. *International Journal of Refractory Metals & Hard Materials*, **25**, 322–327.

114 Hirota, K., Hara, H. and Kato, M. (2007) Mechanical properties of simultaneously synthesized and consolidated carbon

nanofiber (CNF)-dispersed SiC composites by pulsed electric-current pressure sintering. *Materials Science and Engineering A*, **458**, 216–225.

115 Wang, Z., Iqbal, Z. and Mitra, S. (2006) Rapid, low temperature microwave synthesis of novel carbon nanotube-silicon carbide composite. *Carbon*, **44**, 2804–2808.

116 Yamamoto, G., Yokomizo, K., Omori, M., Sato, Y., Jeyadevan, B., Motomiya, K., Hashida, T., Takahashi, T., Okubo, A. and Tohji, K. (2007) Polycarbosilane-derived SiC/single-walled carbon nanotube nanocomposites. *Nanotechnology*, **18**, 1456141–1456145.

117 Katsuda, Y., Gerstel, P., Narayanan, J., Bill, J. and Aldinger, F. (2006) Reinforcement of precursor-derived Si-C-N ceramics with carbon nanotubes. *Journal of the European Ceramic Society*, **26**, 3399–3405.

118 Kroke, E., Li, Y.L., Konetschny, C., Lecomte, E., Fasel, C. and Riedel, R. (2000) Silazane derived ceramics and related materials. *Materials Science & Engineering. R*, **26**, 97–199.

119 Balazsi, C., Konya, Z., Weber, F., Biro, L.P. and Arato, P. (2003) Preparation and characterization of carbon nanotube reinforced silicon nitride composites. *Materials Science and Engineering C*, **36**, 1133–1137.

120 Balazsi, C., Shen, Z., Konya, Z., Kasztovszky, Z., Weber, F., Vertesy, Z., Biro, L.P., Kiricsi, I. and Arato, P. (2005) Processing of carbon nanotube reinforced silicon nitride composites by spark plasma sintering. *Composites Science and Technology*, **65**, 727–733.

6
Physical Properties of Carbon Nanotube–Ceramic Nanocomposites

6.1
Background

Metals are often used as electromagnetic wave shielding materials at radio and microwave frequencies in electronic devices. The high electromagnetic wave shielding arises from their superior electrical conductivity associated with partially filled conduction band structure. The shortcomings of metals for electromagnetic interference (EMI) shielding include their heavy weight and corrosion degradation upon exposure to severe environments. Polymers and ceramics are generally regarded as insulators because of their low electrical and thermal conductivity. To improve the electrical conductivity, conductive fillers are added into polymers or ceramics to form composite materials. Conducting polymer composites are widely studied by many researchers because of their excellent flexibility and superior processability [1, 2]. However, conducting polymer composites can only be used at ambient and mild temperatures due to the low melting temperature of polymers. In contrast, ceramics with high melting temperature, low density, high strength and superior corrosion resistance are being designed for used in advanced electronic and telecommunication industries. For such applications, high-temperature environments are often encountered during their service lives. With this perspective in mind, CNT–ceramic nanocomposites with excellent mechanical, electrical and thermal conducting properties are ideal high-performance materials for applications in extreme conditions such as high temperatures and mechanical stresses [3].

When conductive fillers are introduced into an insulating matrix, its electrical conductivity depends greatly on the concentration and aspect ratio of the fillers. At low filler loading, the electrical conductivity of such composites is relatively low and nearly close to that of the insulating matrix as a result of large interparticle distance (Figure 6.1). The interparticle distance is reduced dramatically when a sufficient amount of filler is added. At a critical filler concentration, the fillers tend to link each other together to form conductive pathways across the insulating matrix. This critical volume concentration of filler is defined as the percolation threshold. Above the critical threshold, the conductivity and dielectric constant of the composites approach those of the filler medium and increase dramatically by several orders of magnitude.

Carbon Nanotube Reinforced Composites: Metal and Ceramic Matrices. Sie Chin Tjong
Copyright © 2009 WILEY-VCH Verlag GmbH & Co. KGaA, Weinheim
ISBN: 978-3-527-40892-4

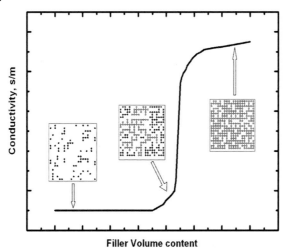

Figure 6.1 Schematic diagram showing electrical behavior and microstructure of conductor-insulator composites with the filler content below, at and above percolation threshold.

In this respect, the composite transforms from an electrical insulator into a conductor. Percolative composites offer attractive applications as switches and sensors because the transition can be externally stimulated by changes in filler concentration, pressure or temperature. The percolative approach needs very low filler content compared with the conventional composite counterpart having a much higher filler concentration.

The electrical behavior of a conductor-insulator composite near the critical filler content can be best described by percolation theory. Classical percolation models (bond and site percolation issues) are established statistically from the random filling of empty sites with filler particles or the random connection of adjacent sites in a filled lattice [4, 5]. Two nearest-neighbor sites are considered to be connected when they are both occupied. Considering the occupied sites are electrical conductors whilst empty sites are insulators, electrons can only flow between the nearest-neighbor conductor sites. The filler volume fraction is determined by the product of the site occupation probability and the filler particle filling factor. The percolation thresholds determined from these models depend on the lattice type (e.g. square, triangular, simple cubic, BCC, FCC), particle coordination number and the filling factor of the particle. In practice, the percolation concentration in real composite materials is more complicated and is generally known to be dependent on the shape and aspect ratio of fillers, dispersion of particles in the matrix and the processing conditions. Fibril fillers with large aspect ratios facilitate formation of conducting path network at lower percolation concentration than spherical fillers [6]. A very low percolation threshold of 0.0025 wt% MWNT has been found in the MWNT/epoxy nanocomposites. This value is far lower than the theoretical prediction of the percolation theory [7].

Recent advancement in microelectronic technology has led to the miniaturization of electronic devices. This results in the escalation of power dissipation and an increase in the heat flux in the devices. The heat dissipation issue is of primarily importance to the performance and reliability of electronic devices. For composite materials, formation of thermally conductive networks through appropriate packing of the fillers in the matrix is an effective route to achieve rapid heat dissipation. Therefore, CNTs with excellent electrical and thermal conductivity are attractive fillers for insulating ceramics that are widely used in electronic industries. Ideally, CNTs with large aspect ratios favor formation of a thermally conductive network near the percolation threshold. However, there is no experimental evidence of thermal percolation threshold in CNT composites [8, 9]. It is generally known that the transport of heat in materials occurs by phonons. When heat is transported across the interface of composites, a temperature discontinuity occurs at the interface due to interfacial resistance. This originates from the acoustic mismatch and weaker interatomic bonding between the filler and matrix at the interface [10].

6.2
Electrical Behavior

The electrical conductivity of CNT–ceramic nanocomposites depends greatly on the processing route employed. Peigney et al. [Chap. 5, Ref. 59] employed a catalytic CVD process for the in situ production of Fe-Al_2O_3/CNT composite powders. Double-walled CNTs or SWNTs were synthesized in situ using metal oxide solid solution precursors. The synthesized composite powders were then subjected to hot pressing at 1500 °C under 43 MPa for 15 min. The electrical conductivity of Fe-Al_2O_3/8.5 vol% CNT and Fe-Al_2O_3/10 vol% CNT nanocomposite was determined to be 40–80 and 280–400 S m^{-1}, respectively. As mentioned before, such in situ nanocomposite is quite porous having a relative density of 88.7%. The high hot-pressing temperature causes a structural damage to CNTs, thereby yielding lower electrical conductivity.

Mukherjee and coworkers [11] prepared Al_2O_3/5.7vol%SWNT, Al_2O_3/10vol% SWNT and Al_2O_3/15vol% SWNT nanocomposites by blending and ball milling SWNTs with alumina nanopowders, followed by spark plasma sintering (SPS) at 1150–1200 °C for 3 min. Pristine alumina was also prepared by SPS for the purpose of comparison. The results are listed in Table 6.1. Apparently, the addition of 5.7 vol% SWNT to alumina increases its electrical conductivity from 10^{-12} to 1050 S m^{-1}, being fifteen orders of magnitude of enhancement in conductivity. The electrical conductivity of alumina increases further to 3345 S m^{-1} by adding 15 vol% SWNT. A dramatic increase is electrical conductivity is attributed to the retention of the integrity of SWNTs and densification of such nanocomposites prepared by SPS at lower temperatures.

Yamamoto et al. [Chap. 5, Ref. 71] fabricated Al_2O_3/0.9vol% MWNT, Al_2O_3/1.9vol% MWNT and Al_2O_3/3.7vol% MWNT nanocomposites by dispersing acid treated MWNTs in ethanol ultrasonically. Aluminum hydroxide and magnesium hydroxide were then added to the nanotube suspension under sonication. The resulting composite powders were subjected to SPS at 1500 °C under a pressure of 20 MPa for 10 min. In spite of the formation of dense nanocomposites, the relatively high sintering

Table 6.1 Effect of sintering temperature on the electrical conductivity of Al$_2$O$_3$/CNT nanocomposites.

Materials	SPS conditions	Relative density (%)	Electrical conductivity (S m^{-1})
Al$_2$O$_3$ Ref [11]	1150 °C, 3 min	100	10^{-12}
Al$_2$O$_3$/5.7vol% SWNT Ref [11]	1150 °C, 3 min	100	1050
Al$_2$O$_3$/10 vol% SWNT Ref [11]	1200 °C, 3 min	99	1510
Al$_2$O$_3$/15 vol% SWNT Ref [11]	1150 °C, 3 min	99	3345
Al$_2$O$_3$ Ref [5.71]	1500 °C, 10 min	98.6	10^{-10}–10^{-12}
Al$_2$O$_3$/0.9vol% MWNT [Chap. 5, Ref. 71]	1500 °C, 10 min	99.2	1.3×10^{-3}
Al$_2$O$_3$/1.9vol% MWNT [Chap. 5, Ref. 71]	1500 °C, 10 min	98.9	1.3
Al$_2$O$_3$/3.7vol% MWNT [Chap. 5, Ref. 71]	1500 °C, 10 min	97.7	65.3

temperature (1500 °C) destroys the structural integrity of CNTs. Consequently, the electrical conductivity of these nanocomposites is rather low (Table 6.1).

6.3
Percolation Concentration

Percolation theory is commonly used to describe and analyze the behavior of connected clusters in a random structure, in order to predict whether a system is macroscopically connected or not. This macroscopic connectivity is of primarily importance to many physical phenomena such as fluid flow, electric current flow, heat flux, diffusion, and so on [4, 5]. Broadbent and Hammersley introduced lattice models for the flow of fluid particles through a random medium. They reported that the fluid would not flow if the concentration of active medium is smaller than a threshold value or percolation probability [12]. For the resistor network medium, the flow of electrons across resistors is analogous to the spread of fluid particles in a random medium. Kirkpatrick extended this concept into a random resistor network consisting of conducting and nonconducting materials [13, 14]. The corresponding expression for d.c. electrical conductivity (σ) is given by:

$$\sigma \propto (P-P_c)^t \tag{6.1}$$

where P and P_c are the occupied probability and percolation probability of the lattice, and t is the conductivity exponent.

In continuum percolation theory, the σ and dielectric constant (ε) of the composite containing conducting fillers generally follow the power law relation [15, 16]:

$$\sigma \propto (p-p_c)^t \quad \text{for} \quad p>p_c \tag{6.2}$$

$$\sigma \propto (p_c-p)^{-s} \quad \text{for} \quad p>p_c \tag{6.3}$$

$$\varepsilon \propto |p-p_c|-s' \quad \text{for} \quad p<p_c, p>p_c \tag{6.4}$$

where p is volume fraction of the filler and p_c the percolation threshold. The critical exponents s, s' and t are assumed to be universal. The conductivity exponent t reflects the dimensionality of the system with values about 1.3 and 2.0 for two and three dimensional random percolation system, respectively [17]. However, the experimental conductivity exponent t value can deviate from the universal value as a result of electron tunneling from conducting fillers above the percolation threshold [18]. According to the literature, the measured percolation threshold in CNT–ceramic composites is very low, typically in the range of 0.64–4.7 vol%, depending on the type of ceramic matrix and processing technology employed [19, 20].

In terms of a.c. properties, the experimental conductivity and dielectric constant display a pronounced frequency-dependent behavior. The frequency dependence of effective conductivity and dielectric constant of a conductor-insulator composite near the percolation threshold can be described by the following universal power law relation [16]:

$$\sigma \propto \upsilon^x \tag{6.5}$$

$$\varepsilon \propto \upsilon^{-y} \tag{6.6}$$

where υ is frequency. The critical exponents x and y must satisfy the following relation:

$$x + y = 1. \tag{6.7}$$

Peigney and coworkers investigated the percolation behavior of *in situ* MgAl$_2$O$_3$/CNT nanocomposites having 0.2–25 vol% CNT [19]. The nanocomposites were prepared by catalytic chemical vapor deposition followed by hot pressing at 1300 °C. Figure 6.2(a) and (b) shows the variation of electrical conductivity with CNT content for the MgAl$_2$O$_3$/CNT nanocomposites. The inset in Figure 6.2(a) represents a log-log plot of the conductivity as a function of p-p_c. From least-square analysis, a linear fit can be obtained according to Equation 6.2, yielding an exponent $t = 1.73 \pm 0.02$. and percolation threshold $p_c = 0.64 \pm 0.02$ vol% (about 0.31 wt%). At the percolation threshold, the electrical conductivity increases over seven orders of magnitude from 10^{-8} to 0.4 S m^{-1}. Comparing with the ultralow value of p_c (0.0025 wt%) for MWNT/epoxy nanocomposites [7], the percolation threshold is two orders of magnitude higher for the MgAl$_2$O$_3$/CNT nanocomposites. This is attributed to the degradation of CNTs during hot-pressing at 1300 °C.

Very recently, Ahmad et al. studied the electrical conductivity and dielectric behavior of Al$_2$O$_3$/MWNT nanocomposites prepared by SPS at 1350 °C [20]. A percolation threshold of 0.79 vol% was found for such nanocomposites. Further, the dielectric constant can reach as high as 5000 in the low frequency region by adding 1.74 vol% MWNT. Shi and Liang [21] also investigated the percolation and dielectric behavior 3Y-TZP (3 mol% yttria-stabilized tetragonal polycrystalline zirconia) filled with various MWNT contents. The 3Y-TZP/MWNT nanocomposites were fabricated by ball milling constituent materials followed by SPS at 1250 °C under a pressure of 60 MPa. Figure 6.3 shows the effective d.c. conductivity vs CNT concentration for the 3Y-TZP/MWNT nanocomposites. Using the least square fit, the

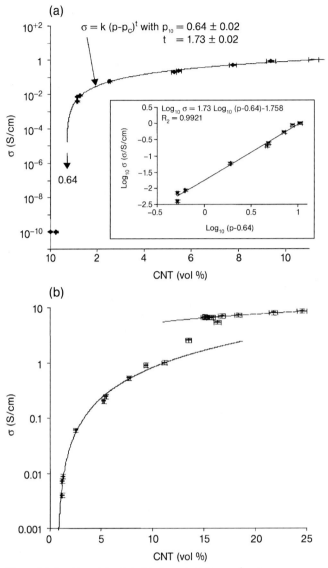

Figure 6.2 Semi-log plots of electrical conductivity as a function of CNT content in the range of (a) 0–11 vol% CNT and (b) 0–25 vol% CNT. Reproduced with permission from [19]. Copyright © (2004) Elsevier.

percolation threshold and conductivity exponent are determined to be 1.7 wt% (4.7 vol%) and 3.30 ± 0.03, respectively. It is noted that the conductivity exponent t is much higher that that predicted by the universal three-dimensional lattice value ($t \approx 2$). They explained this in terms of the thermal fluctuation-induced tunneling effect [21, 22]. As the surface of conducting nanotube fillers is covered with a thin

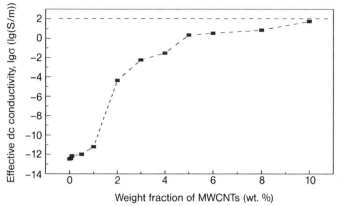

Figure 6.3 Effective d.c. electrical conductivity (σ_{DC}) vs MWNT concentration for 3Y-TZP/MWNT nanocomposites. Reproduced with permission from [21]. Copyright © (2006) John Wiley & Sons, Inc.

zirconia material near the percolation threshold, the charge carrier can be transported across the fillers by tunneling through insulating zirconia. This behavior can be described by the fluctuation induced tunneling model which takes into account tunneling through potential barriers of varying height due to local temperature fluctuations [23]:

$$\sigma \sim \exp[-T_1/(T+T_0)] \tag{6.8}$$

where

$$T_1 = wA\varepsilon^2/8\pi k \tag{6.9}$$

$$T_0 = 2\,T_1/\pi\chi w \tag{6.10}$$

$$\chi = (2\,mV_0/h^2)^{1/2} \tag{6.11}$$

$$\varepsilon = 4\,V_0/ew \tag{6.12}$$

where m and e are electron mass and charge k the Boltzman constant, V_0 the potential barrier height, w the internanotube distance (gap width), and A the area of the capacitance formed by the junction. At constant temperature, Equation 6.8 can be further simplified to [24]:

$$\ln \sigma \sim -w. \tag{6.13}$$

Assuming the nanotube dispersion in the insulating matrix is homogeneous, the composite conductivity at a given temperature can be described by the behavior of a single tunnel junction where the gap width is $w \sim p^{-1/3}$ [25]. Accordingly, the d.c. conductivity should follow the following rule [24]:

$$\ln \sigma \sim -p^{-1/3}. \tag{6.14}$$

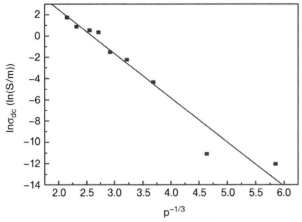

Figure 6.4 Linear variation of ln σ$_{DC}$ versus p$^{-1/3}$. Reproduced with permission from [21]. Copyright © (2006) John Wiley & Sons, Inc.

Figure 6.4 shows the variation of electrical conductivity of the nanocomposites with $p^{-1/3}$. A linear relationship is observed for this plot indicating that tunneling conduction occurs in the nanocomposites.

Figure 6.5 shows the dielectric constant vs filler concentration plot for the 3Y-TZP/MWNT nanocomposites. The dielectric constant also rises abruptly by two orders of magnitude near the percolation threshold. From Equation 6.4, the nanocomposites with p_c of 1.7 wt% yields a critical exponent $s' = 0.63 \pm 0.02$. The conducting nanotube fillers are separated by thin dielectric ceramic layers, forming mini-capacitors across the matrix. The polarization effect between such filler clusters can improve the electric charge storage, thereby contributing to the enhancement of dielectric

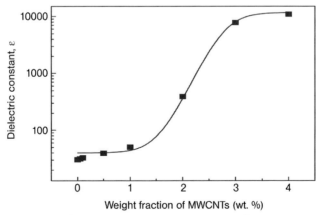

Figure 6.5 Dielectric constant vs MWNT concentration for 3Y-TZP/MWNT nanocomposites. Reproduced with permission from [21]. Copyright © (2006) John Wiley & Sons, Inc.

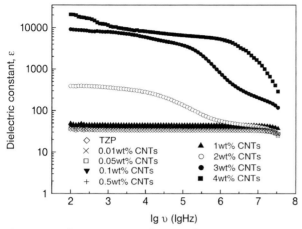

Figure 6.6 Dielectric constant vs frequency for 3Y-TZP/MWNT nanocomposites. Reproduced with permission from [21]. Copyright © (2006) John Wiley & Sons, Inc.

constant of nanocomposites [16]. Figure 6.6 shows the dielectric constant vs frequency plot for the 3Y-TZP/MWNT nanocomposites at room temperature. The dielectric constant of pristine 3Y-TZP and 3Y-TZP/MWNT nanocomposites containing nanotubes up to 1 wt% is quite low and independent of frequency. At MWNT concentration ≥ 2 wt%, the dielectric constant increases significantly. A pronounced dependence of the dielectric constant on frequency is observed for the 3Y-TZP/MWNT nanocomposites with MWNT concentration ≥ 2 wt%. In this regard, the dielectric behavior can be described by Equation 6.6. The dielectric constant can reach 11 200 at 10^3 Hz for the 3Y-TZP/4 wt% MWNT nanocomposite.

6.4
Electromagnetic Interference Shielding

Electromagnetic interference (EMI) shielding has received increasing attention in the electronic and communication industries due to the devices emit radiation at microwave and radio frequencies. The EMI of a material refers to its capability to reflect and absorb electromagnetic radiation at high frequencies. Thus, shielding and absorbing are two main methods to prevent and attenuate EMI radiation. The EMI shielding effectiveness (SE) of a material is expressed as the ratio of transmitted power (E_T) to incident power (E_I) of an electromagnetic wave and measured in decibels (dB) [26, 27]:

$$SE = -10 \log \left(\frac{E_T}{E_I} \right). \tag{6.15}$$

In general, electrical conducting materials such as metals, graphite, conducting polymers and polymer-based composites have been used for EMI shielding applications [28]. Carbon nanotubes with excellent electrical conductivity show attractive applications for EMI shielding. Introducing conductive CNTs into insulating polymers and ceramics is a simple and effective way to increase their EMI effectiveness. The EMI behavior of CNT-polymer nanocomposites is well documented in the literature because they are easy to process [29–31]. In contrast, less information is available on the EMI behavior of CNT–ceramic nanocomposites.

Xiang et al. studied the EMI shielding properties of hot-pressed SiO_2/MWNT nanocomposites in the frequency range of 8–12 GHz (X-band) and 26.5–40 GHz (Ka-band) [32, 33]. They reported that the EMI SE of the nanocomposites increases with increasing nanotube content. Further, the SiO_2/10 vol% MWNT nanocomposite exhibits the highest SE value of 66 dB at 34 GHz. This demonstrates a possible use of such material for commercial applications in the broad-band microwave frequency. The improvement of shielding effectiveness is primarily attributed to improvement of the conductivity (Table 6.2). MWNTs with excellent electrical conductivity and high aspect ratio can easily form conducting networks within a silica matrix. The conducting networks would interact and attenuate the electromagnetic radiation effectively. For the purpose of comparison, silica composites reinforced with carbon black (CB) nanoparticles of 24 nm were also prepared under the same conditions. The EMI SE values of the SiO_2/10 vol% CB composite at 10 and 34 GHz are quite low (i.e., 10–11 dB) due to the low aspect ratio of CB particles.

In addition to EMI SE, another material parameter typically used to characterize the microwave interaction is the complex permittivity (ε^*). The dielectric permittivity of a material is commonly given relative to that of free space, and is known as relative permittivity (ε_r), or dielectric constant. The dielectric responses of the polymer nanocomposites at various frequencies are described in terms of the complex permittivity (ε^*) given by the following equation:

$$\varepsilon^* = \varepsilon' - i\varepsilon'' \tag{6.16}$$

and the dissipation factor (tangent loss) which is defined as:

$$\tan \delta = \varepsilon'' / \varepsilon'. \tag{6.17}$$

Table 6.2 Electrical conductivity and EMI shielding effectiveness of SiO_2/MWNT and SiO_2/CB nanocomposites.

MWNT or CB content (vol%)	Conductivity (S m^{-1})		SE at 10 GHz (dB)		SE at 34 GHz (dB)	
	MWNT composites	CB composites	MWNT composites	CB composites	MWNT composites	CB composites
2.5	6.49×10^{-12}	—	7	—	12	—
5	4	3.17×10^{-12}	22	4	41	10
7.5	21.8	—	26	—	53	—
10	57.4	8.77×10^{-3}	32	10	66	11

Reproduced with permission from [33]. Copyright © (2007) Elsevier.

The real part (ε') correlates with the dielectric permittivity and the imaginary part (ε'') represents the dielectric loss. The tangent loss or dissipation factor represents the ability of the material to convert the absorbed electromagnetic energy into heat [34].

Figure 6.7(a)–(c) show the complex permittivity spectra vs frequency for the SiO_2/MWNT nanocomposites. At 26.5 Hz, the real part of permittivity increases sharply with increasing MWNT content and reaches a value of 21 for the SiO_2/10 vol% MWNT nanocomposite. The ε'' and tan δ spectra display a similar increasing trend with filler content. The ε'' and tan δ values of the SiO_2/10 vol% MWNT nanocomposite at 26.5 Hz approach 28 and 1.3, respectively. The large ε'' and tan δ values give rise to high EM attenuation by absorbing the incident radiation and dissipating it as heat. The dramatic enhancement of complex permittivity can be attributed to the addition of MWNTs with superior electrical conductivity and large aspect ratio. It can be concluded that the EMI properties of insulating ceramics can be tailored effectively by simply adding CNTs having superior load-bearing capacity and EM absorption characteristics.

6.5
Thermal Behavior

By analogy to the electrical properties, the thermal conductivity of ceramics should be markedly improved after the addition of CNTs with very high thermal conductivity. Individual MWNT has unusually high thermal conductivity of 3000 W mK^{-1} [Chap. 1, Ref. 157]. Isolated SWNT has even higher conductivity of 6000 W mK^{-1}. Accordingly, a thermal percolation network is expected to form in the CNT–ceramic nanocomposites as in the case of electrical percolation. This allows for rapid heat flow along the percolating nanotube networks and further enhancement of thermal transport. However, the thermal conductivity of the CNT–ceramic nanocomposites is much smaller than the value predicted from the intrinsic thermal conductivity of the nanotubes and their volume fraction. Further, enhancement in thermal properties is considerably lower than the corresponding improvement in electrical properties [35–37]. This is due to the formation of Kapitza contact resistances at the nanotube–matrix interface [8, 9, 38–41]. It is well recognized that the interface plays an important role in heat conduction through CNT composites. The resistance at the nanotube–ceramic and nanotube–nanotube interfaces restricts heat transport along percolating pathways of CNTs. When a phonon travels along nanotubes and encounters nanotube–ceramic and nanotube–nanotube interfaces, it is blocked and scattered at these sites. On the basis of MD simulations, Huxtable *et al.* demonstrated that heat transport in a nanotube composite material is restricted by the exceptionally small interface thermal conductance, resulting in the thermal conductivity of the composite becomes much lower than the value estimated from the intrinsic thermal conductivity of the nanotubes and their volume fraction [38].

Table 6.3 gives a typical example of the modest improvement in thermal conductivity of silica by adding MWNTs. Dense SiO_2/MWNT nanocomposites were prepared by ball milling of composite slurries followed by SPS treatment at 950–1050 °C

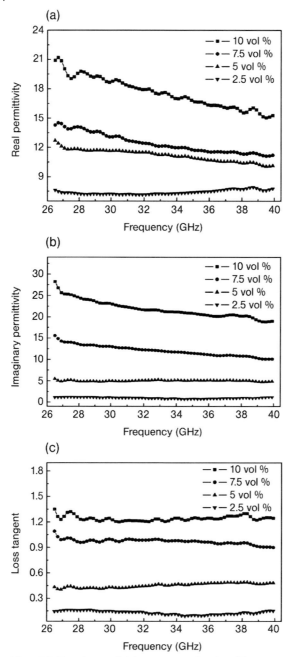

Figure 6.7 Complex permittivity spectra as a function of frequency for SiO_2/MWNT nanocomposites: (a) real permittivity, (b) imaginary permittivity and (c) loss tangent. Reproduced with permission from [33]. Copyright © (2007) Elsevier.

Table 6.3 SPS processing conditions of SiO$_2$/MWNT nanocomposites and their relative densities, grain sizes and thermal conductivities measured at room temperature.

Materials	Processing conditions (°C, MPa, min)	Relative density (%TD)	Grain size (nm)	Thermal conductivity (W m^{-1} K^{-1})
Pure SiO$_2$	950/50/5	100	—	2.42 ± 0.16
Pure SiO$_2$	1050/50/5	100	—	2.47 ± 0.05
SiO$_2$/5vol% MWNT	950/50/5	100	21.08	3.36 ± 0.02
SiO$_2$/5vol% MWNT	950/50/10	100	21.08	3.41 ± 0.02
SiO$_2$/5vol% MWNT	1050/50/5	100	21.92	3.45 ± 0.01
SiO$_2$/5vol% MWNT	1050/50/10	100	21.92	3.48 ± 0.01
SiO$_2$/10vol% MWNT	950/50/5	100	21.08	3.67 ± 0.02
SiO$_2$/10vol% MWNT	950/50/10	100	18.78	3.62 ± 0.05
SiO$_2$/10vol% MWNT	1050/50/5	100	21.08	3.97 ± 0.04
SiO$_2$/10vol% MWNT	1050/50/10	100	22.12	4.08 ± 0.01

Reproduced with permission from [35]. Copyright © (2007) Elsevier.

for 5 and 10 min. A maximum room temperature thermal conductivity of 4.08 W mK^{-1} can be achieved in silica by adding 10 vol% MWNT, corresponding to 65% enhancement. However, the enhancement in conductivity is far smaller than that expected from the unusually high thermal conductivity of CNTs.

Shenogina et al. analyzed the reasons for the lack of thermal percolation in CNTs composites despite the occurrence of electrical percolation [8]. They explained this in terms of a large difference in values between thermal conductivity ratio (C_f/C_m) and electrical conductivity ratio (σ_f/σ_m) of the nanotube to the matrix of composites. This approach can be used to analyze the thermal behavior of SiO$_2$/MWNT nanocomposites. Assuming $C_f = 3000$ W m^{-1} K^{-1}, $C_m = 2.47$ W m^{-1} K^{-1}, thus $C_f/C_m = 1215$. Pure silica is an insulator with high electrical resistivity of ~10^{10} Ω m. The reported room temperature resistivity of individual CNT is in the order of 10^{-8}–10^{-6} Ω m [Chap. 1, Ref. 164]. As electrical conductivity equals to the inverse of electrical resistivity, we obtain $\sigma_f/\sigma_m = 10^{18}$–$10^{16}$. The large difference between σ_f/σ_m and C_f/C_m explains the absence of thermal percolation in the SiO$_2$/MWNT nanocomposites.

Nomenclature

A	Area of the capacitance
C	Thermal conductivity
e	electron charge
E_I	Electromagnetic wave incident power
E_T	Electromagnetic wave transmitted power
k	Boltzman constant
m	electron mass

p	Filler volume fraction
p_c	Percolation threshold
P	Occupied probability of the lattice
P_C	Percolation probability of the lattice
s'	Dielectric exponent
SE	Shielding effectiveness
t	Conductivity exponent
$\tan \delta$	Tangent loss or dissipation factor
V_0	Potential barrier height
w	Internanotube distance
x	Critical exponent
y	Critical exponent
ε	Dielectric constant
ε'	Real permittivity
ε''	Imaginary permittivity
ε^*	Complex permittivity
σ	Electrical conductivity
υ	Frequency

References

1 Mamunya, Y.P., Davydenko, V.V., Pissis, P. and Lebedev, E.V. (2002) Electrical and thermal conductivity of polymers filled with metal powders. *European Polymer Journal*, **38**, 1887–1897.

2 Psarras, G.C., Manolakaki, E. and Tsangaris, G.M. (2002) Electrical relaxations in polymeric particulate composites of epoxy resin and metal particles. *Composites A*, **33**, 375–384.

3 Zhan, G., Kuntz, J.D. and Mukherjee, A.K. (2005) Ceramic materials reinforced with single-wall carbon nanotubes as electrical conductors. US Patent 6875374.

4 Stauffer, D. and Aharony, A. (1992) *Introduction to Percolation Theory*, 2nd edn, Taylor & Francis, London.

5 Sahimi, M. (1994) *Applications of Percolation Theory*, Taylor & Francis, London.

6 Celzard, A., McRae, E., Deleuze, C., Dufort, M., Furdin, J. and Mareche, J.F. (1996) Critical concentration in percolating systems containing a high-aspect-ratio filler. *Physical Review B-Condensed Matter*, **53**, 6209–6214.

7 Sandler, J.K., Jirk, J.E., Kinloch, I.A., Shaffer, M.S. and Windle, A.H. (2003) Ultra-low electrical percolation threshold in carbon nanotube-epoxy composites. *Polymer*, **44**, 5893–5899.

8 Shenogina, N., Shenogin, S., Xue, L. and Keblinski, P. (2005) On the lack of thermal percolation in carbon nanotubes composites. *Applied Physics Letters*, **87**, 1331061–1331063.

9 Biercuk, M.J., Liaguno, M.C., Radosavljevic, M., Hyun, J.K., Johnson, A.T. and Fischer, J.E. (2002) Carbon nanotube composites for thermal management. *Applied Physics Letters*, **80**, 2767–2769.

10 Pettersson, S. and Mahan, G.D. (1990) Theory of the thermal boundary resistance between dissimilar lattices. *Physical Review B-Condensed Matter*, **42**, 7386–7390.

11 Zhan, G.D., Kuntz, J.D., Garay, J.E. and Mukherjee, A.K. (2003) Electrical properties of nanoceramics reinforced

with ropes of single-walled carbon nanotubes. *Applied Physics Letters*, **83**, 1228–1230.
12. Broadbent, S.R. and Hammersley, J.M. (1957) Percolation process, I: Crystals and mazes. *Proceedings of the Cambridge Philosophical Society*, **53**, 629–641.
13. Kirkpatrick, S. (1973) Percolation and conduction. *Reviews of Modern Physics*, **45**, 574–588.
14. Kirkpatrick, S. (1976) Percolation phenomena in higher dimensions: Approach to the mean-field limit. *Physical Review Letters*, **36**, 69–72.
15. Webman, I., Jortner, J. and Cohen, M.H. (1977) Critical exponents for percolation conductivity in resistor networks. *Physical Review B-Condensed Matter*, **16**, 2593–2596.
16. Song, Y., Noh, T.W., Lee, S.I. and Gaines, J.R. (1986) Experimental study of the three-dimensional ac conductivity and dielectric constant of a conductor-insulator composite near the percolation threshold. *Physical Review B-Condensed Matter*, **33**, 904–909.
17. Clerc, J.P., Giraud, G., Laugier, J.M. and Luck, J.M. (1990) The electrical conductivity of binary disordered systems, percolation clusters, fractals and related models. *Advances in Physics*, **39**, 191–209.
18. Balberg, I. (1987) Tunneling and nonuniversal conductivity in composite materials. *Physical Review Letters*, **59**, 1305–1308.
19. Rul, S., Lefevre-schlick, F., Capria, E., Laurent, C. and Peigney, A. (2004) Percolation of single-walled carbon nanotubes in ceramic matrix nanocomposites. *Acta Materialia*, **52**, 1061–1067.
20. Ahmad, K., Pan, W. and Shi, S.L. (2006) Electrical conductivity and dielectric properties of multiwalled carbon nanotube and alumina composites. *Applied Physics Letters*, **89**, 133122(1)–133122(3).
21. Shi, S.L. and Liang, J. (2006) Effect of multiwall carbon nanotubes on electrical and dielectric properties of yttria-stabilized zirconia ceramic. *Journal of the American Ceramic Society*, **89**, 3533–3535.
22. Shi, S.L. and Liang, J. (2007) Electronic transport properties of multiwall carbon nanotubes/yttria stabilized zirconis composites. *Journal of Applied Physics*, **101**, 023708(1)–023708(5).
23. Sheng, P., Sichel, E.K. and Gittleman, J.L. (1978) Fluctuation-induced tunneling in carbon-polyvinylchloride composites. *Physical Review Letters*, **40**, 1197–1200.
24. Ezquerra, T.A., Kulescza, M. and Balta-Calleja, F.J. (1991) Electrical transport in polyethylene-graphite composite materials. *Synthetic Metals*, **41**, 915–920.
25. Connor, M.T., Roy, S., Ezquerra, T.A. and Balta-Calleja, F.J. (1998) Broadband ac conductivity of conductor-polymer composites. *Physical Review B-Condensed Matter*, **57**, 2286–2294.
26. Ott, H.W. (1987) *Noise Reduction Techniques in Electronic Systems*, John Wiley & Sons, Inc., New York.
27. Joo, J. and Lee, C.Y. (2000) High frequency electromagnetic interference shielding response of mixtures and multilayer films based on conducting polymers. *Journal of Applied Physics*, **88**, 513–518.
28. Chung, D.D.L. (2001) Electromagnetic interference shielding effectiveness of carbon materials. *Carbon*, **39**, 279–285.
29. Yang, Y., Gupta, M.C., Dudley, K.L. and Lawrence, R.W. (2005) a comparative study of EMI properties of carbon nanofiber and multi-walled carbon nanotube filled polymer composites. *Journal of Nanoscience and Nanotechnology*, **5**, 927–931.
30. Li, N., Huang, Y., Du, F., He, X., Lin, X., Gao, H. and Ma, Y. (2006) Electromagnetic interference (EMI) shielding of single-walled carbon nanotube epoxy composites. *Nano Letters*, **6**, 1141–1145.
31. Yuan, S.M., Ma, C.C., Chuang, C.Y., Yu, K.C., Wu, S.Y., Yang, C.C. and We, M.H. (2008) Effect of processing method on the shielding effectiveness of electromagnetic

interference of MWCNT/PMMA composites. *Composites Science and Technology*, **68**, 963–968.

32 Xiang, C., Pan, Y., Liu, X., Sun, X., Shi, X. and Guo, J. (2005) Microwave attenuation of multiwalled carbon nanotube-fused silica composites. *Applied Physics Letters*, **87**, 123103(1)–123103(3).

33 Xiang, C., Pan, Y. and Guo, J. (2007) Electromagnetic interference shielding effectiveness of multiwalled carbon nanotube reinforced fused silica composites. *Ceramics International*, **33**, 1293–1297.

34 Atwater, J.E. and Wheeler, R.R. (2004) Microwave permittivity and dielectric relaxation of a high surface area activated carbon. *Applied Physics A-Materials Science & Processing*, **79**, 125–129.

35 Sivakumar, R., Guo, S., Nishimura, T. and Kagawa, Y. (2007) Thermal conductivity in multi-wall carbon nanotube/silica-based nanocomposites. *Scripta Mater*, **56**, 265–268.

36 Jiang, L. and Gao, L. (2008) Densified multiwalled carbon nanotubes-titanium nitride composites with enhanced thermal properties. *Ceramics International*, **34**, 231–235.

37 Kumari, L., Zhang, T., Du, G.H., Li, W.G., Wang, Q.W., Datye, A. and Wu, K. (2008) Thermal properties of CNT-alumina nanocomposites. *Composites Science and Technology*, **68**, 2178–2183.

38 Huxtable, S., Cahill, D.G., Shenogin, S., Xue, L., Ozisik, R., Barone, P., Usrey, M., Strano, M.S., Siddons, G., Shim, M. and Keblinski, P. (2003) Interfacial heat flow in carbon nanotubes suspensions. *Nature Materials*, **2**, 731–734.

39 Nan, C.W., Liu, G., Lin, Y. and Li, M. (2004) Interface effect on thermal conductivity of carbon nanotube composites. *Applied Physics Letters*, **85**, 3549–3551.

40 Ju, S. and Li, Z.Y. (2006) Theory of thermal conductance in carbon nanotube composites. *Physics Letters A*, **353**, 194–197.

41 Xu, Q.Z. (2006) Model for the effective thermal conductivity of carbon nanotube composites. *Nanotechnology*, **17**, 1655–1660.

7
Mechanical Properties of Carbon Nanotube–Ceramic Nanocomposites

7.1
Fracture Toughness

Mechanical strength, hardness and fracture toughness are key parameters in the materials selection process for ceramic materials. Ceramics are attractive structural materials for engineering applications because of their high mechanical strength. However, ceramics are brittle as a result of their high resistance to dislocation slip. Designing to prevent catastrophic failure of ceramic materials under the application of external stresses requires a fundamental knowledge of their fracture mechanisms.

Fracture toughness is fundamental design property of materials containing cracks that undergo fracture as a result of unstable crack propagation. A fracture mechanics approach is developed to assess the materials' resistance to fracture in the presence of a crack. Linear elastic fracture mechanics (LEFM) is used to evaluate the toughness of brittle materials that fracture in elastic deformation regime. In this approach, stress intensity factor (K) is used to characterize the magnitude of stress field at a crack tip during mechanical loading. The critical value of the stress intensity factor resulting from unstable crack propagation and final failure of the materials under tensile mode is termed as "fracture toughness" or K_{IC}. Most ceramic materials suffer fracture before the onset of plastic deformation. The fracture toughness of ceramics can be determined experimentally using different specimen configurations, such as single edge precracked beam (SEPB), single edge notched (SENB), single edge V-notched beam (SEVNB) and chevron notched beam (CNB). Such specimens are mechanically loaded under three-point or four-point flexural conditions [1–6]. These measurements can give bending strength and true fracture toughness values that can be accurately reproduced. Further, SEPB, SEVNB and CNB measurements have been accepted as standard test methods by the American Society for Testing and Materials (ASTM). Generally, a pre-existing sharp crack must be made in specimens prior to the fracture toughness measurements. For instance, SEPB specimens are pre-cracked in a specially designed bridge–anvil tool under a compressive load. This yields quite a large crack tip radius over the entire width. Thus, the geometry of the crack is difficult to control. For SENB specimens, the pre-crack can be simply made by using a diamond

Carbon Nanotube Reinforced Composites: Metal and Ceramic Matrices. Sie Chin Tjong
Copyright © 2009 WILEY-VCH Verlag GmbH & Co. KGaA, Weinheim
ISBN: 978-3-527-40892-4

saw or machining. The notch-root radius must be smaller than 10 μm in order to obtain reliable fracture toughness results [7]. With the SEVNB it is relatively easy to sharpen the V-notch with a razor blade. In general, ceramic materials are rather sensitive to the sharpness of the crack tip. A large notch-root tends to overestimate the real fracture toughness of ceramics. For instance, the SEVNB fracture toughness of alumina with a notch-root radius of ~300 μm is quite large, that is, 7.6 MPa m$^{1/2}$. This toughness can be reduced to an acceptable value of 3.8 MPa m$^{1/2}$ by using a notch-root radius of <10 μm [8].

For the SENB specimen under three-point bending, the stress intensity factor can be determined from the following equation [9]:

$$K_{IQ} = Y \frac{3PS\sqrt{a}}{2BW^2} \quad (7.1)$$

$$Y = 1.93 - 3.07(a/W) + 14.53(a/W)^2 - 25.11(a/W)^3 + 25.8(a/W)^4 \quad (7.2)$$

where Y is a geometric factor, P the load at failure, S the span length, a is the crack length, B and W are the thickness and width of the specimen, respectively. K_{IQ} becomes K_{IC} when the plane strain and linear elastic failure criteria are met. For the SEVNB specimen under four-point bending, the stress intensity factor can be evaluated using the following relation [10]:

$$K_{IQ} = Y \frac{P}{B\sqrt{W}} \frac{S_1 - S_2}{W} \frac{3\sqrt{\alpha}}{2(1-\alpha)^{1.5}}. \quad (7.3)$$

$$Y = 1.9887 - 1.326\alpha - (3.49 - 0.68\alpha + 1.35\alpha^2)\alpha(1-\alpha)(1+\alpha)^{-2} \quad (7.4)$$

where α is given by $\alpha = (a_N + a)/W$; and a_N is the notch length, S_1 and S_2 are the outer and inner loading spans, respectively.

For simplicity, Vickers indentation fracture test (VIF) is employed by many researchers to determine the fracture toughness of ceramics and their composites [11]. A polished specimen is indented by a Vickers pyramidal microhardness indenter during which deformation takes place below the indentation. Cracks emanating from four corners of the impression in the deformation zone are propagated to the material surface. The fracture toughness can be evaluated from the radial crack length (*c*) measuring from the indentation center, the indentation load (*P*), the elastic modulus (*E*) and hardness (*H*) of the tested material by the Antis' equation [12,13]:

$$K_{IC} = 0.016 \left(\frac{E}{H}\right)^{1/2} \left(\frac{P}{c^{3/2}}\right) \quad (7.5)$$

Thus, the crack length at a given applied load determines the toughness of test material. Values of fracture toughness can also be evaluated by using other empirical equations such as Niihara [14,15] and Miyoshi [16]. No standard tests specimens are needed to evaluate the toughness value.

Very recently, Quinn from National Institute of Standards and Technology indicated that VIF cannot be used to characterize the fracture toughness of ceramic materials [17]. This is because VIF result cannot be reproduced, hence does not yield a reliable K_{IC} value. He compared the K_{IC} values of silicon nitride specimens obtained from standardized fracture toughness test and VIF technique. He found that Antis' equation underestimates the fracture toughness whilst Niihara's equation overestimates the toughness. Although Miyoshi equation yields a result close to that of standardized fracture toughness test, a calibration constant is needed to fit the data. Therefore, the fracture toughness of brittle ceramics should be determined from the standardized fracture toughness test. The standard fracture test produces reliable fracture toughness values because the specimen with a well-defined single crack is subjected to a well-defined mechanical loading condition. On the contrary, VIF test has a complex three-dimensional cracks and ill-defined crack arrest condition. For new ceramic materials with unknown fracture toughness value, VIF test should not be used to determine the toughness. This is because VIF toughness can deviate up to 48% from the true fracture toughness value [18]. Despite these deficiencies, the VIF test is used extensively to determine the fracture toughness of ceramic-CNT nanocomposites.

7.2
Toughening and Strengthening Mechanisms

The mechanisms responsible for toughening of CNT-reinforced ceramics include crack deflection, crack bridging and nanotube pull-out. The toughening behavior can be explored from the Vickers indentation of ceramic-CNT nanocomposites synthesized from AAO membranes. Such nanocomposite coatings contain a highly ordered array of parallel nanotubes in an alumina matrix. Xia *et al.* prepared highly aligned alumina-MWNT nanocomposites using AAO coatings of different thicknesses (20 and 90 μm) through template synthesis [Chap. 5, Ref. 81]. After multi-step anodizing, Co or Ni catalyst particles were deposited into the bottom of nanopores of amorphous alumina coatings. CVD treatment at 645 °C was employed to grow MWNTs up the pore walls. Controlling the composition of reactant gases and deposition time produces MWNTs with different wall thickness. The nanotubes formed inside the AAO coating with 90 μm thickness have a larger diameter and a thinner wall thickness than those formed with 20 μm. The nanocomposites were then subjected to nanoindentation using a Berkovitch indenter. Interestingly, alumina-MWNT nanocomposites exhibit all the toughening features similar to those observed in conventional fiber-reinforced composites, that is, crack deflection, crack bridging and crack pull-out (Figure 7.1(a)–(c)). Consequently, interface debonding and sliding can occur in composite materials with nanostructures. Furthermore, the nanocomposite having nanotubes with thinner wall thickness and larger diameter is more resistant to indentation damage. This is demonstrated by the collapse of CNTs into shear bands, leading to the absorption of large energy during indentation. No racking is observed during deformation of the nanocomposite coating (Figure 7.2). The shear band formation is somewhat similar to shear yielding observed in polymers. For polymeric materials, shear banding often initiates from

(a)

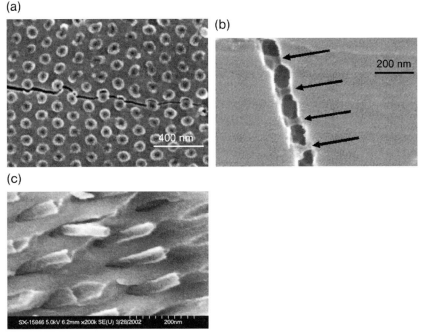

(b)

(c)

Figure 7.1 SEM micrographs showing (a) crack intersection with the successive alumina/MWNT interfaces and defection around the MWNTs along the interface; (b) CNT bridging the gap between the crack surfaces; (c) nanotube pull-out. Reproduced with permission from [Chap. 5, Ref. 81]. Copyright © (2004) Elsevier.

cavitation of second-phase elastomer particles followed by the shear yielding of the polymer matrix. Consequently, much energy can be dissipated during mechanical deformation, thereby improving the fracture toughness of polymers considerably.

For highly disordered ceramic-CNT nanocomposites, crack deflection, crack bridging and crack pull-out mechanisms are often observed [Chap. 5, Ref. 44,

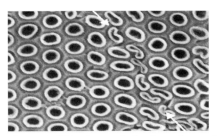

Figure 7.2 SEM micrograph showing collapse of the nanotubes into a shear band (indicated by the arrows) for the nanocomposite coating of 90 μm thickness. Reproduced with permission from [Chap. 5, Ref. 81]. Copyright © (2004) Elsevier.

Chap. 5, Ref. 54, Chap. 5, Ref. 60]. The strengthening of such nanocomposites arises from effective load transfer from the matrix to nanotubes during mechanical loading. Homogeneous dispersion of CNTs in ceramics and strong nanotube–matrix bonding are prerequisites to achieve improved mechanical strength in the nanocomposites.

7.3
Oxide-Based Nanocomposites

7.3.1
Alumina Matrix

7.3.1.1 Deformation Behavior

The mechanical properties of alumina-CNT nanocomposites depend strongly on the composite synthesis route and sintering process. The composite synthesis route controls the distribution of CNTs in the ceramic matrix and the interfacial bonding states. The later sintering process determines the final density of nanocomposite products. In general, homogeneous dispersion of nanotubes and strong CNT-matrix interfacial bonding are difficult to achieve using conventional powder mixing techniques. Ball milling of ceramic powders and CNTs in ethanol is essential to assist homogeneous dispersion but it beneficial effects depends greatly on the milling time, ball-powder ratio and type of ceramic powders. The heterocoagulation route shows promise because the alumina-CNT composite with better nanotube dispersion and strong interfacial bonding can be synthesized at the molecular level.

Conventional hot pressing that combines composite powder consolidation and heat treatment at high temperatures is often adopted by researchers to fabricate nanocomposites due to its simplicity and versatility. Care must be taken in the selection of hot-pressing temperatures in order to avoid damage to the nanotubes at elevated temperatures. Peigney et al. [Chap. 5, Ref. 59] reported that the fracture strength of hot-pressed Fe-Al_2O_3/4.8 wt% CNT nanocomposite is only marginally higher than that of monolithic alumina, but lower than that of the Fe-Al_2O_3 composites. Further, the fracture toughness values of Fe-Al_2O_3/8.5 vol% CNT and Fe-Al_2O_3/10 vol% CNT nanocomposites were lower than that of Fe-Al_2O_3 (Table 7.1).

Table 7.1 Mechanical properties of hot-pressed alumina and alumina-CNT nanocomposites.

Materials	Fe content (wt%)	Fracture strength (MPa)	Fracture toughness (MPa m$^{1/2}$)	Fracture mode
Al_2O_3 Ref [Chap. 5, Ref. 42]	—	330	4.4	Intergranular
Fe-Al_2O_3 Ref [Chap. 5, Ref. 42]	10	630	7.2	Mixed
Fe-Al_2O_3/8.5 vol%CNT Ref [Chap. 5, Ref. 59]	8.38	400	5.0	Mixed
Fe-Al_2O_3/10 vol% CNT Ref [Chap. 5, Ref. 59]	8.38	296	3.1	Intergranular

These nanocomposites were prepared by hot pressing of synthesized *in situ* composite powders at 1500 °C. The relative density (r.d.) values of hot-pressed Fe-Al_2O_3/8.5 vol% CNT and Fe-Al_2O_3/10 vol% CNT composites are 88.7 and 87.3%, respectively. Therefore, the alumina-CNT nanocomposites exhibit inferior mechanical properties due to their porous nature.

More recently, Maensiri *et al.* investigated the mechanical properties of alumina/CNF nanocomposites reinforced with 1, 2.5 and 5 vol% CNF [19]. The nanocomposites were prepared by conventional powder mixing, ball-milling, followed by hot pressing at 1450 °C. From SEM fractographs as shown in Figure 7.3(a)–(d)), agglomeration of CNFs can be readily seen in the Al_2O_3/5 vol% CNF composite. The bending stress of hot-pressed Al_2O_3, Al_2O_3/1 vol% CNF, Al_2O_3/2.5 vol% CNF and Al_2O_3/5 vol% CNF specimens is 414, 301, 279 and 219 MPa, respectively. Therefore, the bending strength of hot-pressed nanocomposites decreases with increasing filler content as a result of filler clustering and degradation of CNFs at high pressing temperature. The VIF toughness of hot-pressed Al_2O_3, Al_2O_3/1 vol% CNF, Al_2O_3/2.5 vol% CNF and Al_2O_3/5 vol% CNF specimens is 2.41, 2.50, 2.74 and 2.54 MPa m$^{1/2}$, respectively. The Al_2O_3/2.5 vol% CNF nanocomposite exhibits the highest fracture toughness, but it is only improved by 13.7% compared with monolithic alumina. The toughening mechanism in Al_2O_3/2.5 vol% CNF nanocomposite is associated with the crack bridging effect of CNF (Figure 7.3(c)).

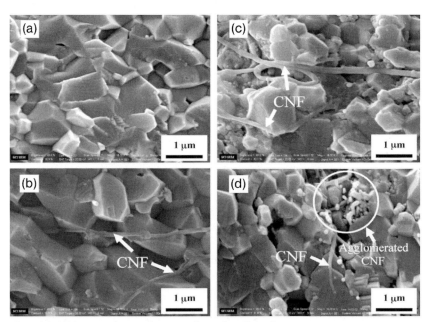

Figure 7.3 SEM images showing fracture surfaces of (a) Al_2O_3; (b) Al_2O_3/1 vol% CNF; (c) Al_2O_3/2.5 vol% CNF; (d) Al_2O_3/5 vol% CNF specimens. Reproduced with permission from [19]. Copyright © (2007) Elsevier.

Fan et al. prepared the $Al_2O_3/0.5\,wt\%$ SWNT and $Al_2O_3/1\,wt\%$ SWNT nanocomposites using heterocoagulation [Chap. 5, Ref. 54]. The colloidally-dispersed composite and monolithic alumina powders were hot pressed at 1600 °C in an argon atmosphere under a pressure of 20 MPa for 1 h. The r.d. values of monolithic alumina, $Al_2O_3/0.5\,wt\%$ SWNT and $Al_2O_3/1\,wt\%$ SWNT nanocomposites are 99.8%, 99.5% and 99.1%. The bending strength of Al_2O_3, $Al_2O_3/0.5\,wt\%$ SWNT and $Al_2O_3/1\,wt\%$ SWNT nanocomposites is 356, 402 and 423 MPa, respectively. The VIF toughness of Al_2O_3, $Al_2O_3/0.5\,wt\%$ SWNT and $Al_2O_3/1\,wt\%$ SWNT nanocomposites is 3.16, 5.12 and 6.40 MPa m$^{1/2}$. Addition of 1 wt% SWNT improves the fracture toughness of alumina by 103% and bending strength by 18.8%. Crack bridging and pull-out of SWNTs can be readily seen in the SEM fractographs of $Al_2O_3/1\,wt\%$ SWNT nanocomposite and is responsible for the toughening behavior of this material. It is noted that the mechanical properties of the $Al_2O_3/0.5\,wt\%$ SWNT and $Al_2O_3/1\,wt\%$ SWNT nanocomposites can be improved further by hot pressing the composite powders at lower temperatures. Hot pressing the nanocomposite at 1600 °C leads to the degradation of SWNTs. It has been reported that that the structure of CNTs of alumina nanocomposites can be preserved by employing lower sinter temperatures up to ~1250 °C only. Above this temperature, CNTs convert to graphite and carbon nano-onion structures [Chap. 5, Ref. 53, Chap. 5, Ref. 59].

Sun et al. investigated the effect of hot-pressing temperatures on the bending strength and fracture toughness of $Al_2O_3/1\,wt\%$ MWNT nanocomposite [Chap. 5, Ref. 52]. The nanocomposite was prepared by the heterocoagulation route and hot pressing. The later mechanical treatment was carried out at 1350–1500 °C in an argon atmosphere under 30 MPa for 1 h. Figures 7.4 and 7.5 show SEM

Figure 7.4 (a) Low and (b) high magnification SEM micrographs showing the fracture surface of $Al_2O_3/1\,wt\%$ MWNT nanocomposite sintered at 1500 °C. Reproduced with permission from [Chap. 5, Ref. 52]. Copyright © (2005) Elsevier.

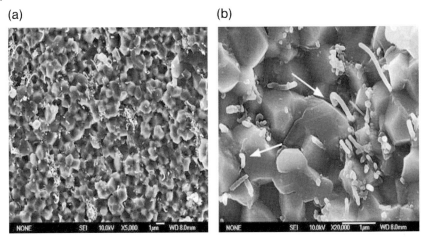

Figure 7.5 (a) Low and (b) high magnification SEM micrographs showing the fracture surface of Al$_2$O$_3$/1 wt% MWNT nanocomposite sintered at 1450 °C. Reproduced with permission from [Chap. 5, Ref. 52]. Copyright © (2005) Elsevier.

fractographs of the Al$_2$O$_3$/1 wt% MWNT nanocomposite sintered at 1500 °C and 1450 °C, respectively. More pores (indicated by arrows) and fewer MWNTs are found in the nanocomposite sintered at 1500 °C. The r.d. of this nanocomposite sintered at 1500 and 1450 °C is 97.9 and 98.7%, respectively. The mechanical properties of this composite hot pressed at different temperatures are summarized in Table 7.2. Hot pressing at 1500 °C results in lowest VIF toughness (2.2 MPa m$^{1/2}$) and bending stress (200 MPa). The lower r.d. and fewer MWNTs for the specimen

Table 7.2 Room temperature fracture toughness and bending strength of monolithic alumina and Al$_2$O$_3$/1 wt% MWNT.

Sintering temperature (°C)	Fracture toughness (MPa m$^{1/2}$)	Bending strength (MPa)
Al$_2$O$_3$/1 wt% MWNT		
1350	4.0 ± 0.50	536 ± 44
1400	3.7 ± 0.41	550 ± 32
1450	3.9 ± 0.59	554 ± 64
1500	2.2 ± 0.15	200 ± 49
Al$_2$O$_3$		
1300	3.7 ± 0.93	496 ± 64
1350	3 ± 0.28	383 ± 97
1400	3.7 ± 0.45	460 ± 73
1450	3.9 ± 0.50	388 ± 29
1500	3.7 ± 0.34	371 ± 16

Reproduced with permission from [Chap. 5, Ref. 52]. Copyright © (2005) Elsevier.

sintered at 1500 °C are responsible for its poorest fracture toughness and lowest bending stress. Reducing the sinter temperature to 1450 °C leads to a marked increase of the bending stress (554 MPa) and VIF toughness (3.9 MPa m$^{1/2}$). The fracture toughness of nanocomposite increases with decreasing sinter temperature. Comparing the fracture toughness values of monolithic alumina sintered at the same temperature, it appears that the toughening efficiency of MWNT in alumina is very low. The fracture toughness of nanocomposite hot pressed at 1500 °C is even smaller than that of alumina sintered at 1500 °C. There is no improvement in fracture toughness of the nanocomposite hot pressed at 1450 and 1400 °C. The fracture toughness of nanocomposite hot pressed at 1350 °C improves by nearly 33.3% compared with the single-phase alumina sintered at 1350 °C. The poor toughness of the nanocomposite hot pressed at 1400–1500 °C is attributed to the degradation of CNTs [Chap. 5, Ref. 53, Chap. 5, Ref. 69]. It is quite obvious that the hot-pressing temperature must be reduced to lower temperatures to preserve the structure of CNTs.

In another study, Sun et al. employed SPS to consolidate the Al_2O_3/0.1 wt% CNT nanocomposite and single-phase alumina at 1300 °C for 5 min [Chap. 5, Ref. 61]. Obviously, the addition of only 0.1 wt% CNT to alumina increases the fracture toughness by about 32.4% from 3.7 to 4.9 MPa m$^{1/2}$. Compared with hot pressing, only one-tenth of CNT content is needed to improve the toughness of alumina by ∼33%. Figure 7.6(a) and (b) show SEM fractographs of Al_2O_3/0.1 wt% CNT nanocomposite. It is interesting to see from Figure 7.6(a) that the nanotubes are covered with matrix material, forming bridges across the crack. The crack bridging effect hinders the enlargement of the crack. Other alumina-sheathed nanotubes are broken and embedded in the matrix along the crack (Figure 7.4(b)). The crack bridging mechanism is responsible for the improvement of fracture toughness of such nanocomposite. This behavior is commonly observed in the polymer-CNT nanocomposites toughened by CNTs [20]. The polymer sheathed nanotubes bridge the matrix cracks effectively during tensile loading (Figure 7.7).

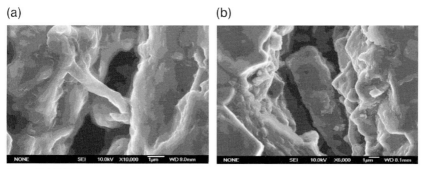

Figure 7.6 SEM fractographs showing typical interactions between crack and alumina sheathed CNTs. Reproduced with permission from [Chap. 5, Ref. 61]. Copyright © (2002) The American Chemical Society.

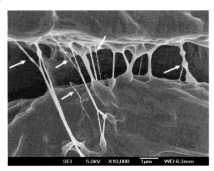

Figure 7.7 SEM micrograph showing crack bridging by polymer sheathed CNTs. Some nanotubes are sheathed by small polymer beads as indicated by the arrows. Reproduced with permission from [20]. Copyright © (2004) The American Chemical Society.

Yamamoto et al. fabricated the Al_2O_3/0.9 vol% MWNT, Al_2O_3/1.9 vol% MWNT and Al_2O_3/3.7 vol% MWNT nanocomposites by dispersing acid-treated MWNTs in ethanol ultrasonically. Aluminum hydroxide and magnesium hydroxide were then added to the nanotube suspension under sonication. The resulting suspension was filtered, oven dried, followed by SPS at 1500 °C under a pressure of 20 MPa for 10 min [Chap. 5, Ref. 71]. From the zeta potential measurements, aluminum hydroxide is positively charged over a wide pH range of 3–9 whilst acid-treated MWNTs is negatively changed in this pH range. Aluminum hydroxide particles are accumulated on MWNTs surfaces due to the electrostatic attraction, resulting in a homogeneous dispersion of nanotubes. Aluminum hydroxide precursor finally converted to α-alumina during SPS. The VIF test revealed that the addition of 0.9 vol% MWNT to alumina leads to an increase of fracture toughness from about 4.25 to 5.90 MPa m$^{1/2}$. The toughness then decreases with increasing nanotube content. Figure 7.8 shows the variation of normalized fracture toughness vs nanotube content for the Al_2O_3/MWNT nanocomposites. In this figure, the fracture toughness value of nanocomposites is normalized to that of pure alumina. The normalized experimental result reported by Fan et al. [Chap. 5, Ref. 54] is also shown for comparison. Apparently, additions of CNTs up to 1 vol% CNTs can produce a beneficial toughening effect in alumina despite the employment of high sintering temperature (1500 or 1600 °C). Above 1 vol% CNT, the toughening effect of CNTs diminishes as expected.

Mukerjee and coworkers fabricated pure Al_2O_3, Al_2O_3/5.7 vol% SWNT and Al_2O_3/10 vol% SWNT nanocomposites through powder mixing in ethanol, ball milling and SPS at 1150 °C for 3 min under 63 MPa [Chap. 5, Ref. 44, Chap. 5, Ref. 70]. The fracture toughness and hardness of these specimens were determined by indentation measurements. The indentation fracture toughness of pure Al_2O_3, Al_2O_3/5.7 vol% SWNT and Al_2O_3/10 vol% SWNT nanocomposites is 3.3, 7.9 and 9.7 MPa m$^{1/2}$, respectively. All these specimens achieved a full density of 100% by sintering at 1150 °C for 3 min (Chap. 5, Table 5.2). The structure and unique properties of SWNTs can be preserved at 1150 °C for 3 min. Under these processing

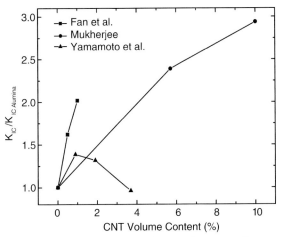

Figure 7.8 Normalized Vickers indentation fracture toughness values vs nanotube content for Al_2O_3/CNT nanocomposites sintering at different temperatures.

conditions, the fracture toughness of dense Al_2O_3/10 vol% SWNT nanocomposite is nearly three times that of pure alumina. The normalized fracture toughness values of the nanocomposites as a function of filler content is also plotted in Figure 7.8. The beneficial toughening effect of SWNTs can be readily seen in this figure. Figure 7.9(a)–(c) show SEM fractographs of the Al_2O_3/10 vol% SWNT nanocomposite. Crack bridging, crack deflection and nanotube pull-out can be readily seen in an indentation crack of this nanocomposite. These mechanisms are responsible for the improved fracture toughness of the nanocomposite.

The enhanced indentation fracture toughness of Al_2O_3/CNT nanocomposites associated with CNT additions has raised a doubt over the reliability of K_{IC} values obtained from Vickers indentation tests. Recently, there has been a serious discussion and debate on the toughening effect of adding SWNT to alumina determined by Vickers indentation [21–23]. Padture and coworkers re-examined the effect of SWNT addition on the fracture toughness of the Al_2O_3/10 vol% SWNT nanocomposite. Both VIF and SEVNB standard test methods were used to evaluate the fracture toughness [21]. They reported that CNT has little effect in improving the fracture toughness of alumina. The Al_2O_3/10 vol% SWNT nanocomposite was prepared by powder mixing in methanol, ball milling and SPS at 1450–1550 °C for 2–3 min under a pressure of 40 MPa. For the purpose of comparison, Al_2O_3/10 vol% graphite composite was also fabricated under the same processing conditions. Further, dense and porous alumina specimens with grain sizes of ~1–2 μm were prepared by cold isostatic pressing of the green compact and pressureless sintering at 1600 °C for 30 min or 1350 °C for 60 min. The r.d. of dense alumina, porous alumina, Al_2O_3/10 vol% SWNT nanocomposite and Al_2O_3/10 vol% graphite composite is 98.1, 90.6, 95.1 and 97.7%, respectively.

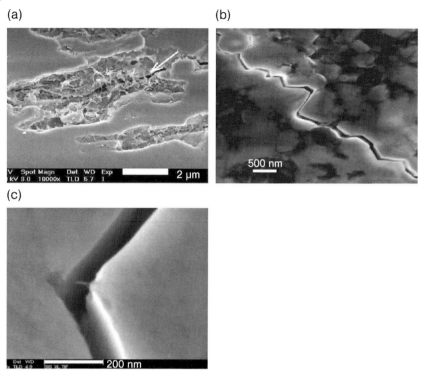

Figure 7.9 (a) Crack bridging; (b) crack deflection; (c) nanotube pull-out, observed in an indentation crack of the alumina/10 vol% SWNT nanocomposite. Reproduced with permission from [Chap. 5, Ref. 70]. Copyright © (2007) Elsevier.

Figure 7.10(a) shows low and high-magnification SEM micrographs of the dense alumina subjected to VIF test. Radial cracks can be readily seen in these micrographs, and the toughness is determined to be 3.01 MPa m$^{1/2}$ using the Antis' equation. A similar radial cracking behavior can be observed in the porous alumina subjected to Vickers indentation (Figure 7.8(b)). However, classical radial cracks are absent in the Al_2O_3/10 vol% SWNT composite (Figure 7.10(c)). Consequently, the Antis' equation becomes invalid for fracture toughness determination. It is considered that the highly shear-deformable SWNTs can assist redistribution of the stress field under indentation, endowing the nanocomposite with contact-damage resistance. Thus, it appears that the VIF test cannot be used to characterize the fracture toughness of alumina-CNT nanocomposites, but the test results can provide useful information relating the resistance of such materials to indentation fracture.

Direct fracture toughness measurements using the SEVNB specimens under four-point bending yield an average K_{IC} of 3.22 MPa m$^{1/2}$ for the dense alumina, 3.32 MPa m$^{1/2}$ for the Al_2O_3/10 vol% SWNT nanocomposite and 3.51 MPa m$^{1/2}$ for the Al_2O_3/10 vol% graphite composite. These results indicate that there is little

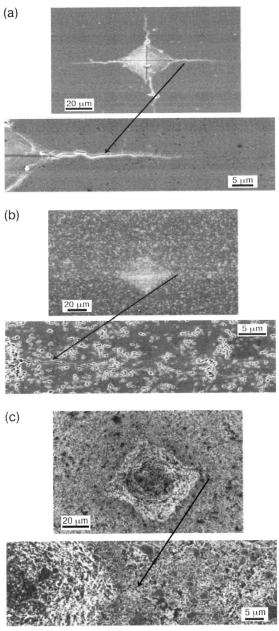

Figure 7.10 Low-and high magnification SEM micrographs of (a) dense and (b) porous alumina showing top views of Vickers indentation sites with radial cracks formation. Low-and high magnification SEM micrographs of (c) Al_2O_3/10 vol% SWNT composite showing no radial cracks formation at indentation sites. Reproduced with permission from [21]. Copyright © (2004) Nature Publishing Group.

Table 7.3 Comparison of processing conditions for making Al$_2$O$_3$/10 vol% SWNT nanocomposite.

Comparison parameters	Mukherjee's group Ref. [Chap. 5, Ref. 44]	Padture's group Ref. [21]
Sintering materials	SWNT and Al$_2$O$_3$	SWNT and Al$_2$O$_3$
Dispersing and mixing methods	Wet-milling and sieving	No sieving
SPS processing conditions	1150 °C, 3 min, 63 MPa	1450–1550 °C, 3–10 min, 40 MPa
Relative density (%)	100	95.1
Grain size	~200 nm	1000–2000 nm
Fracture toughness	194% increase (VIF)	No toughening (VIF and SEVNB)

Reproduced with permission from [23]. Copyright © (2008) Elsevier.

improvement in the facture toughness of alumina by adding 10 vol% SWNT. In other words, the Al$_2$O$_3$/10 vol% SWNT and Al$_2$O$_3$/10 vol% graphite composites are as brittle as the dense alumina. One possible explanation for low fracture toughness of the Al$_2$O$_3$/10 vol% SWNT nanocomposite obtained from the SEVNB standard method is the employment of high SPS temperatures (i.e., 1450–1550 °C) for the consolidation of powder mixture. Table 7.3 summarizes the processing conditions from Mukherjee and Padture's research groups for fabrication of the Al$_2$O$_3$/10 vol% SWNT nanocomposite. Apparently, the SPS temperature of the Al$_2$O$_3$/10 vol% SWNT nanocomposite prepared by Padture's group is significantly higher, thereby leading to large matrix grain size of the resulting composite.

The effect of hybrid reinforcements (1 vol% SiC and 5–10 vol% MWNT) on the mechanical properties of alumina nanocomposites is now considered. The hybrid reinforcements offer distinct advantages by combining different properties of nanofillers to produce nanocomposites with higher toughness. The hybrids were fabricated by blending MWNTs, SiC and alumina in ethanol ultrasonically followed by ball-milling and drying. Dried composite powders were spark plasma sintered at 1550 °C. For comparison, pure alumina and Al$_2$O$_3$/1 vol% SiC nanocomposite were also prepared under the same processing conditions. Figure 7.11 shows the r.d., fracture strength, bending strength and hardness for dense Al$_2$O$_3$, Al$_2$O$_3$/1 vol% SiC composite and Al$_2$O$_3$/(MWNT + SiC) hybrids [Chap. 5, Ref. 75]. It is obvious that SiC nanoparticles inhibit densification of alumina but its addition improves both the bending strength and fracture toughness of alumina. Additions of 5 and 7 vol% MWNT to the Al$_2$O$_3$/1 vol% SiC nanocomposite further improve the fracture toughness but the bending strength decrease slightly. The r.d. and all mechanical properties of the hybrid deteriorate by adding 10 vol% MWNT due to the nanotube agglomeration. SEM fractographs reveal that crack bridging and crack deflection by MWNTs are responsible for improving the toughness of hybrid nanocomposites.

In the case of plasma-sprayed nanocomposite coatings, Agarwal's group indicated that the MWNT additions are beneficial to enhance the toughness of alumina

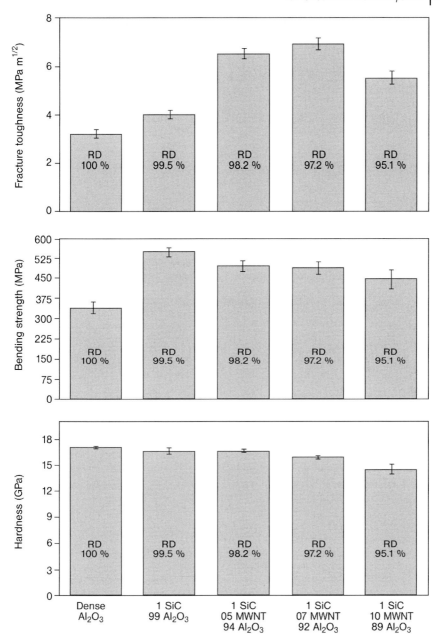

Figure 7.11 Relative density (r.d.) and mechanical properties of monolithic Al_2O_3, Al_2O_3/1 vol% SiC composite and Al_2O_3/(MWNT + SiC) hybrids. Reproduced with permission from [Chap. 5, Ref. 75]. Copyright © (2008) Elsevier.

[Chap. 5, Ref. 76, Chap. 5, Ref. 77]. The fracture toughness of alumina (A-SD) coating is 3.22 ± 0.22 MPa m$^{1/2}$ and increases to 3.862 ± 0.16 MPa m$^{1/2}$ by adding 4 wt% MWNT (A4C-B) coating. The fracture toughness increases to 4.60 ± 0.27 MPa m$^{1/2}$ in the A4C-SD coating due to improved MWNT dispersion. It is further improved to 5.04 ± 0.58 MPa m$^{1/2}$ by adding 8 wt% MWNT (A8C-SD). The enhanced toughness of the alumina/MWNT nanocomposite coating is attributed to the CNT bridge formation, crack deflection and MWNT pull-out [24].

7.3.2
Silica Matrix

Ning et al. fabricated the SiO$_2$/MWNT nanocomposites by direct powder mixing in ethanol and hot pressing at 1300 °C under 25 MPa in nitrogen atmosphere for 30 min [Chap. 5, Ref. 87]. As conventional powder mixing and hot pressing were used, the dispersion of MWNTs in silica matrix is inhomogeneous (Figure 7.12(a) and (b)). Agglomeration of nanotubes is more apparent at 10 vol% MWNT. Thus, the bending strength and fracture toughness of silica can improve by adding only 5 vol% MWNT. These mechanical properties degrade markedly with further increasing nanotube content (Figure 7.13). In another study, Ning et al. fabricated the SiO$_2$/5 vol% MWNT nanocomposite by the sol-gel method with and without C$_{16}$TAB surfactant. The resulting powder mixtures were also hot pressed at 1300 °C under 25 MPa in nitrogen atmosphere [Chap. 5, Ref. 89]. They reported that the bending strength and fracture toughness of the SiO$_2$/5 vol% MWNT nanocomposite are 70.2 ± 5 MPa and 2.18 ± 0.15 MPa m$^{1/2}$, respectively. And the bending strength and fracture toughness of the SiO$_2$/(5 vol% MWNT + C$_{16}$TAB) nanocomposite are 97.0 ± 10 MPa and 2.46 ± 0.11 MPa m$^{1/2}$, respectively. Compared with the mechanical properties of the SiO$_2$/5 vol% MWNT nanocomposite prepared by conventional powder mixing as shown in Figure 7.13, sol-gel prepared SiO$_2$/(5 vol% MWNT + C$_{16}$TAB) nanocomposite exhibits higher bending stress and toughness due to better dispersion of CNTs in silica matrix.

Guo et al. [25] employed colloidal dispersion followed by attrition milling and SPS to produce the SiO$_2$/5 vol% MWNT and SiO$_2$/10 vol% MWNT nanocomposites.

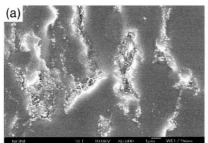
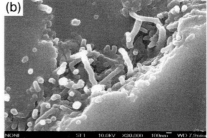

Figure 7.12 (a) Low and (b) high magnification SEM fractographs of SiO$_2$/5 vol% MWNT nanocomposite. Reproduced with permission from [Chap. 5, Ref. 87]. Copyright © (2003) Elsevier.

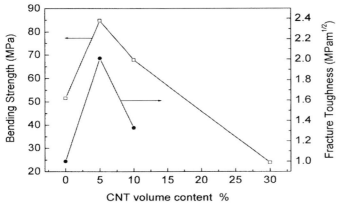

Figure 7.13 Variations of bending strength and fracture toughness with CNTs content for SiO_2/MWNT nanocomposites. Reproduced with permission from [Chap. 5, Ref. 87]. Copyright © (2003) Elsevier.

The SPS treatments were carried out at 950–1050 °C under 50 MPa for 5 or 10 min. As expected, low sintering temperatures facilitate formation of dense monolithic silica and its composites with a r.d. of 100%. The damage to CNTs can be prevented at these temperatures. The indentation K_{IC} value of dense silica is 1.06 MPa m$^{1/2}$. The K_{IC} values of the SiO_2/5 vol% MWNT and SiO_2/10 vol% MWNT nanocomposites sintered at 950 °C for 5 min are 2.14 and 2.68 MPa m$^{1/2}$, respectively. These correspond to 101.9% and 153% increments in fracture toughness over monolithic silica. Carbon nanotube bridging mechanism is responsible for improvement of the fracture toughness. Increasing the sintering temperature to 1050 °C has a small effect on the fracture toughness of the nanocomposites.

7.4
Carbide-Based Nanocomposites

Conventional powder mixing and high temperature sintering only produces a modest improvement in the mechanical properties of SiC-CNT nanocomposites. Ma et al. hot pressed the SiC and nanotube powder mixture at 2000 °C [Chap. 5, Ref. 41]. Without B_4C sintering aid, the resulting nanocomposite is porous having a low r.d. of 64%. Consequently, the bending strength and SENB-type fracture toughness of this specimen are very poor as expected (Table 7.4). The r.d. of the (SiC + 1%B_4C)/10%CNT increases significantly to 94.7% by adding B_4C. However, only slight improvements in bending strength (9.8%) and fracture toughness (10.66%) over monolithic silicon carbide are observed. Hot pressing the (SiC + 1%B_4C)/10%CNT nanocomposite at 2200 °C results in a higher r.d. of 98.1%, but the bending strength reduces dramatically to 227.8 MPa due to the nanotube

Table 7.4 Mechanical properties of SiC + 1%B$_4$C/10% CNT nanocomposite hot pressed at different temperatures.

Materials	Hot-pressing temperature (°C)	Relative density (%)	Bending strength (MPa)	Fracture toughness (MPa m$^{1/2}$)
SiC + 1%B$_4$C	2000	93.9	317.5^a	3.47^a
SiC + 1%B$_4$C/10% CNT	2000	94.7	348.5	3.84^a
SiC + 1%B$_4$C/10% CNT	2200	98.1	227.8	—

Reproduced with permission from [Chap. 5, Ref. 41]. Copyright © (1998) Springer.
^{a}Average value.

degradation. This leads to ineffective load transfer across the nanotube-matrix interface.

As mentioned before, the SPS technique allows consolidation of ceramics and its composites at lower temperatures. Owing to the covalent bonding nature of SiC, the SPS temperature must be controlled at temperatures $\geq 1800\,°C$ to achieve denser microstructure [Chap. 5, Ref. 114]. Figure 7.14 shows the effect of SPS temperature on the bending strength, hardness and indentation fracture toughness of monolithic SiC. Both the bending strength and fracture toughness of monolithic SiC reached an apparent maximum at 1800 °C. Figure 7.15 shows the effect of VGCF addition on the mechanical properties of SiC/VGCF nanocomposites. Apparently, carbon nanofiber additions have no effect in improving the fracture toughness of nanocomposites.

To improve the interfacial bonding between the reinforcement and SiC matrix, CNTs have been coated with SiC layer upon exposure to SiO(g) and CO(g) [Chap. 5, Ref. 48]. Morisada *et al.* [Chap. 5, Ref. 113] studied the effect of SiC-coated MWNTs on the Vickers hardness and indentation fracture toughness of the SiC/MWNT nanocomposites. Figure 7.16 shows the Vickers microhardness vs nanotube content for the SiC nanocomposites reinforced with pristine and SiC-coated nanotubes. Apparently, coating the nanotubes with SiC layer improves the hardness of nanocomposites considerably, thereby facilitating effective load transfer across the nanotube-matrix interface. Further, the indentation fracture toughness of SiC-coated nanocomposites increases with increasing nanotube content (Figure 7.17). The nanocomposite with 3 vol% coated MWNT exhibits the highest indentation toughness of 5.5 MPa m$^{1/2}$. However, this only corresponds to $\sim 14.5\%$ increment over monolithic SiC.

The mixing of CNTs with liquid polymer precursors allows nanotubes to be homogeneously distributed and the following low processing temperature excludes the damage of nanotubes. An *et al.* synthesized Si-C-N/MWNT nanocomposites by cross-linking and pressure-assisted pyrolysis of mixtures containing polyurea(methylvinyl) silazane and MWNTs [Chap. 5, Ref. 56]. They reported that the stiffness of Si-C-N ceramic increases markedly by adding 1.3 and 6.4 vol.% MWNTs. The elastic modulus values of the Si-C-N/MWNT nanocomposites deviate positively from those predicted from the Halpin-Tsai equation. A large deviation of

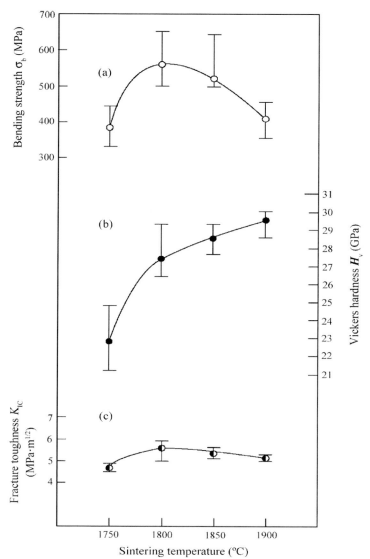

Figure 7.14 (a) Three-point bending strength, (b) Vickers hardness and (c) indentation fracture toughness vs sintering temperature for monolithic SiC prepared by spark plasma sintering. Reproduced with permission from [Chap. 5, Ref. 114]. Copyright © (2007) Elsevier.

elastic modulus from the Halpin-Tsai equation is possibly resulted from the residual stress due to the large shrinkage of nanocomposites during pyrolysis. Katsuda et al. determined the fracture toughness the Si-C-N/MWNT nanocomposites prepared by casting of a mixture of MWNTs and polyureasilazane followed by

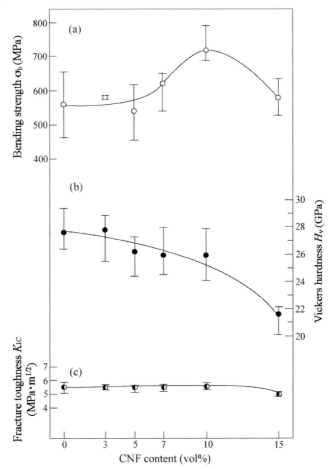

Figure 7.15 (a) Three-point bending strength, (b) Vickers hardness and (c) indentation fracture toughness vs carbon nanofiber content for SiC/VGCF nanocomposites prepared by SPS at 1800 °C under 40 MPa. Reproduced with permission from [Chap. 5, Ref. 114]. Copyright © (2007) Elsevier.

crosss linking and pyrolysis [Chap. 5, Ref. 117]. The thermal loading technique with an edge-cracked circular disk was used to evaluate the fracture toughness of nanocomposites. Two types of MWNTs were used to reinforce the Si-C-N matrix, that is, type-A (high aspect ratio) and type-B (low aspect ratio). Figure 7.18 shows the fracture toughness vs nanotube content for nanocomposites investigated. It is apparent that the fracture toughness of the Si-C-N/MWNT nanocomposites improves significantly by adding type-A nanotubes. There is more than 60% improvement in fracture toughness of Si-C-N ceramic by adding 2 mass% MWNTs. Cracking bridging and nanotube pull-out can be readily seen in the fracture surface of the nanocomposite with 2 mass% MWNT. Little improvement

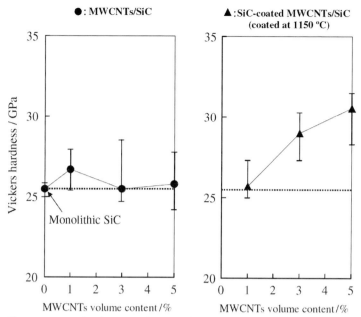

Figure 7.16 Microhardness as a function of CNT content for (a) SiC/MWNT and (b) SiC/SiC-coated MWNT nanocomposites. Reproduced with permission from [Chap. 5, Ref. 113]. Copyright © (2007) Elsevier.

in fracture toughness of the nanocomposites by adding nanotubes with low aspect ratio (Type-B) as expected.

7.5
Nitride-Based Nanocomposites

At present, only Si_3N_4/1% MWNT nanocomposite has been fabricated by conventional powder mixing and ball milling followed by either hipping or SPS treatment at elevated temperatures [Chap. 5, Ref. 43, Chap. 5, Ref. 119, Chap. 5, Ref. 120]. In general, ball milling of composite raw materials in ethanol does not yield adequate dispersion of nanotubes in silicon nitride matrix. In combination with high temperature consolidation, the beneficial effect of nanotube addition on mechanical properties of the Si_3N_4/1% MWNT diminishes. Table 7.5 lists the mechanical properties of monolithic Si_3N_4 and Si_3N_4/1 wt% MWNT nanocomposite prepared by SPS. The hardness values of nanocomposite sintered at 1500 and 1650 °C are lower than those of monolithic silicon nitride sintered at these temperatures. The fracture toughness of nanocomposite spark plasma sintered at 1500 °C is comparable to that of monolithic Si_3N_4 fabricated at the same temperature. However, the toughness of nanocomposite sintered at 1650 °C is inferior to that of Si_3N_4 fabricated at the same temperature.

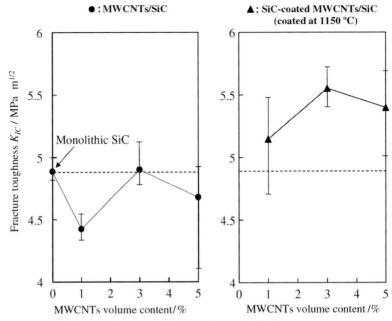

Figure 7.17 Indentation fracture toughness s a function of CNT content for (a) SiC/MWNT and (b) SiC/SiC-coated MWNT nanocomposites. Reproduced with permission from [Chap. 5, Ref. 113]. Copyright © (2007) Elsevier.

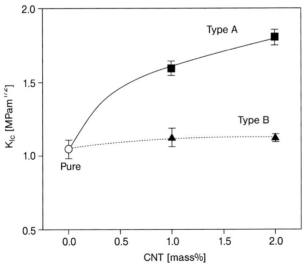

Figure 7.18 Fracture toughness vs CNT content for Si-C-N/MWNT nanocomposites. Reproduced with permission from [Chap. 5, Ref. 117]. Copyright © (2006) Elsevier.

Table 7.5 Mechanical properties of monolithic Si_3N_4 and $Si_3N_4/1$ wt% MWNT nanocomposite spark plasma sintered at different temperatures.

Materials	Sintering temperature (°C)	Density (g cm^{-3})	Vickers hardness (GPa)	Fracture toughness (MPa m$^{1/2}$)
Si_3N_4	1500	3.23	20.1 ± 0.9	5.2
Si_3N_4	1650	3.24	18.3 ± 0.5	6.5
Si_3N_4/1 wt% MWNT	1500	3.17	16.6 ± 0.4	5.3
Si_3N_4/1 wt% MWNT	1650	3.19	19.1 ± 0.6	4.4

Reproduced with permission from [Chap. 5, Ref. 120]. Copyright © (2005) Elsevier.

7.6
Wear Behavior

Hard and superhard ceramic nanocomposite coatings have attracted increasing attention due to their potential applications in diverse areas such as cutting tools, bearings, micro-electro-mechanical systems (MEMS), magnetic disk drives, and so on. Hard coatings improve the durability of substrate materials in hostile environments against severe wear, thus prolonging the materials' life. Typical ceramic nanocomposite coatings are composed of nanocrystalline transition metal carbides embedded in amorphous covalent nitride (e.g., Si_3N_4, or BN) matrix [26]. Hard nanocomposite coatings are often very brittle, which strongly limits their practical use. Tough nanocomposite coatings can be realized by embedding solid lubricant soft phase such as transition metal dichalgonenides (MoS_2, WS_2, $NbSe_2$, etc.) in an amorphous ceramic matrix [27] as shown in Figure 7.19(a) and (b).

As mentioned before, CNTs also exhibit self-lubrication behavior. In combination with their other attractive mechanical and physical properties, CNTs have consistently outperformed their transition metal dichalgonenide rivals. Ceramic-CNT nanocomposites have been shown to exhibit low friction coefficient and wear rate, making them useful as wear-resistant materials for structural components in industrial sectors. In general, the distribution of CNTs in ceramic matrix affects the wear behavior of ceramic-CNT nanocomposites considerably. Recently, Lim *et al.* investigated the effect of CNT dispersion on tribological behavior of the of Al_2O_3/MWNT nanocomposites [28]. Two processing routes were adopted to prepare nanocomposites. The first processing route consisted of ball milling of CNTs and alumina powders in ethanol followed by hot pressing at 1850 °C. The second route was ball milling of slurry consisting of a mixture of MWNT, alumina, methyl-isobutyl ketone, poly(vinyl)butyral and dibutyl phthalate. These organic agents acted as solvent, dispersant, binder and plasticizer, respectively. The slurry was poured into a tape-casting equipment, followed by lamination and hot pressing at 1850 °C. Tape casting gives better distribution of CNTs in alumina matrix, resulting in the formation of dense nanocomposites (Figure 7.20). However, the r.d. of hot-pressed nanocomposites decreases continuously with increasing nanotube content.

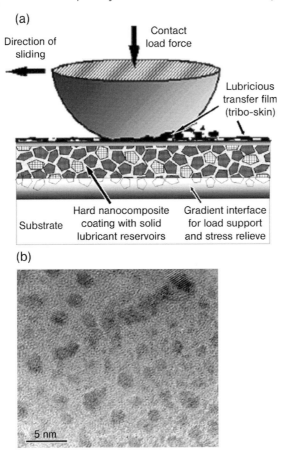

Figure 7.19 (a) Schematic representation of a tough nanocomposite coating design, combining a nanocrystalline/amorphous structure with a functionally gradient interface and (b) TEM image of an Al_2O_3/MoS_2 nanocomposite coating consisted of an amorphous Al_2O_3 ceramic matrix encapsulating 5–10 nm inclusions of nanocrystalline MoS_2 grains. Reproduced with permission from [27]. Copyright © (2005) Elsevier.

Enhanced densification in tape-cast nanocomposites contributes to higher wear resistance and low friction coefficient, as expected. Figures 7.21 and 7.22 show the variations of wear weight loss and friction coefficient with CNT content for Al_2O_3/MWNT nanocomposites prepared by hot pressing and tape casting, respectively. Apparently, the weight loss of the hot-pressed nanocomposites decreases initially by adding 4 wt% MWNT. It then increases with further increasing filler content. This is due to poor dispersion of MWNTs in alumina matrix at higher filler content. In contrast, the weight loss of tape-cast nanocomposites decreases linearly with increasing nanotube content as a result of better dispersion of nanotubes in

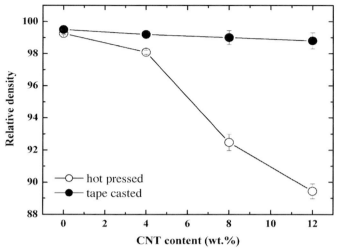

Figure 7.20 Variation of r.d. with CNT content for Al_2O_3/MWNT nanocomposites prepared by hot pressing and tape casting. Reproduced with permission from [28]. Copyright © (2005) Elsevier.

alumina matrix. In another study, Lim and coworkers [29] also demonstrated that the hot-pressed Al_2O_3/CNT nanocomposites prepared by catalytic pyrolysis of acetylene gas with an iron nitrate impreganated alumina exhibit similar wear behavior to that of the Al_2O_3/MWNT nanocomposites as shown in Figure 7.21. A

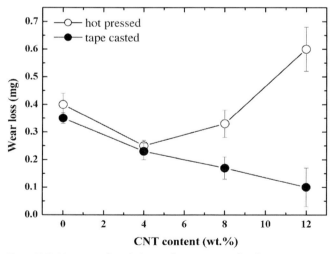

Figure 7.21 Variation of weight loss with CNT content for Al_2O_3/MWNT nanocomposites prepared by hot pressing and tape casting. Reproduced with permission from [28]. Copyright © (2005) Elsevier.

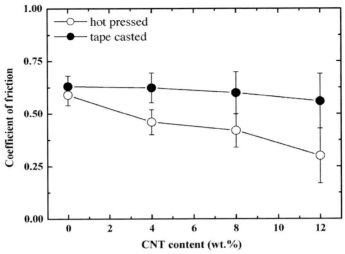

Figure 7.22 Variation of friction coefficient with CNT content for Al$_2$O$_3$/MWNT nanocomposites prepared by hot pressing and tape casting. Reproduced with permission from [28]. Copyright © (2005) Elsevier.

minimum wear loss is also observed at 4 wt% CNT; the weight loss then increases with increasing filler content. They attributed this to the low densification of *in situ* Al$_2$O$_3$/CNT nanocomposites at higher filler contents.

Very recently, Xia *et al.* [30] studied the dry sliding wear behavior of well-aligned Al$_2$O$_3$/MWNT nanocomposites at macro-, micro- and nanoscales by means of the pin-on-disk, microscratch and atomic force microscopy. Aligned Al$_2$O$_3$/MWNT nanocomposites were synthesized through template synthesis using AAO coatings of different thicknesses. Both thin-walled (5 nm) and thick-walled (12 nm) nanotubes were synthesized inside the nanopores accordingly. They reported that the frictional coefficient of the nanocomposites tested at macro-, micro- and nanoscales depends on the buckling behavior of the nanotubes and applied loading. Figure 7.23 shows the coefficient of friction vs sliding distance for porous alumina matrix and aligned Al$_2$O$_3$/MWNT nanocomposites with thin- and thick-walled CNTs at macroscale determined from the pin-on-disk test. The coefficient of friction of porous alumina matrix is about 0.9 and remains nearly constant during testing. The coefficient of friction of thin-walled nanocomposite is slightly lower than that of alumina matrix. However, the thick-wall nanocomposite exhibits a very low coefficient of friction (0.2) at the onset of sliding. The coefficient of friction then increases with the sliding distance but still remains lower than that of the thin-walled counterpart. Figure 7.24 shows the coefficient of friction vs scratching distance for porous alumina matrix and aligned Al$_2$O$_3$/MWNT nanocomposites for the three materials subjected to the microscratch test. The alumina matrix and thin-walled nanocomposite have nearly the same coefficient of friction while the thick-walled nanocomposite displays

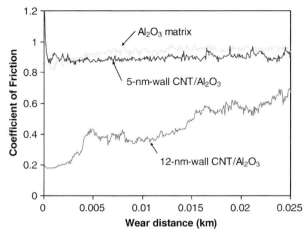

Figure 7.23 Coefficient of friction vs sliding distance for amorphous alumina matrix, thin- and thick-walled Al$_2$O$_3$/MWNT nanocomposites, tested with the pin-on-disk method. Reproduced with permission from [30]. Copyright © (2008) Elsevier.

lower friction at low loading. Figure 7.25 shows the frictional force vs normal force at the nanoscale. The coefficients of friction for porous alumina matrix, thin- and thick-walled nanocomposites are determined to be 0.147, 0.142 and 0.073, respectively. The thick-walled nanocomposite again exhibits the lowest coefficient of friction at the

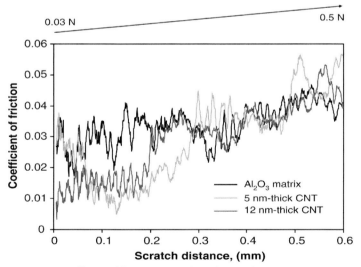

Figure 7.24 Coefficient of friction vs scratching distance for amorphous alumina matrix, thin- and thick-walled Al$_2$O$_3$/MWNT nanocomposites. Reproduced with permission from [30]. Copyright © (2008) Elsevier.

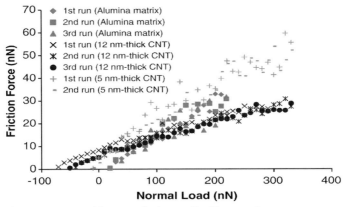

Figure 7.25 Frictional force vs apllied normal load for amorphous alumina matrix, thin- and thick-walled Al$_2$O$_3$/MWNT nanocomposites, measured with atomic force microscopy. Reproduced with permission from [30]. Copyright © (2008) Elsevier.

nanoscale. The difference in the wear behavior of thin- and thick-walled nanocomposites is considered to be associated with the unique contact condition at the frictional interface and the buckling behavior of nanotubes. For the thin-walled nanocomposite, the nanotubes buckle readily, thus the sliding ball tip of wear tester at macroscale can directly contact with the matrix. The multiasperity contacting mode results in a high coefficient of friction. On the other hand, the thick-walled nanotubes are stiffer and resist buckling, thus can mechanically support the matrix during the wear test, provided that the applied loads are not too high.

References

1 Newman, J.C. (1984) A review of Chevron-notched fracture specimens, in *Chevron-notched specimens: testing and stress analysis (ASTM STP 855)* (eds J.H. Underwood, W.E. Freiman and F.I. Baratta), American Society for Testing and Materials, Philadephia, PA, pp. 5–31.

2 Gogotsi, G.A. (2002) Fracture toughness studies on ceramics and ceramic particulate composites at different temperatures, in *Fracture Resistance Testing of Monolithic and Composite Brittle Materials (ASTM STP 1409)* (eds J.A. Salem, G.D. Quinn and M.G. Jenkins), American Society for Testing and Materials, West Conshohocken, PA, pp. 199–212.

3 ASTM C1421-99 (1999) *Standard Test Method for the Determination of Fracture Toughness of Advanced Ceramics at Ambient Temperature*, American Society for Testing and Materials, Philadelphia, PA.

4 Gogotsi, G.A. (2003) Fracture toughness of ceramics and ceramic composites. *Ceramics International*, **29**, 777–784.

5 Mukhopadhyay, A.K., Datta, S.K. and Chakraborty, D. (1999) Fracture toughness of structural ceramics. *Ceramics International*, **25**, 447–454.

6. Fischer, H., Waindich, A. and Telle, R. (2008) Influence of preparation of ceramic SEVNB specimens on fracture toughness testing results. *Dental Materials*, **24**, 618–622.
7. Picard, D., Lequillon, D. and Putot, C. (2006) A method to estimate the influence of the notch-root radius on the fracture toughness measurement of ceramics. *Journal of the European Ceramic Society*, **26**, 1421–1427.
8. Nishida, T., Hanaki, Y. and Pezzotti, G. (1994) Effect of notch-root radius on the fracture toughness of a fine-grained alumina. *Journal of the American Ceramic Society*, **77**, 606–608.
9. Hertzberg, R.W. (1989) *Deformation and Fracture Mechanics of Engineering Materials*, Wiley, New York.
10. Kubler, J. (1997) Fracture toughness of ceramics using the SEVNB method: Preliminary results. *Ceramic Engineering and Science Proceedings*, **18**, 155–162.
11. Sakai, M. and Bradt, R.C. (1993) Fracture toughness testing of brittle materials. *International Materials Reviews*, **38**, 53–78.
12. Antis, G.R., Chantikul, P., Lawn, B.R. and Marshall, D.B. (1981) A critical evaluation of indentation techniques for measuring fracture toughness: I., Direct crack measurements. *Journal of the American Ceramic Society*, **64**, 533–538.
13. Chantikul, P., Antis, G.R., Lawn, B.R. and Marshall, D.B. (1981) A critical evaluation of indentation techniques for measuring fracture toughness: II, Strength methods. *Journal of the American Ceramic Society*, **64**, 539–543.
14. Niihara, K. (1983) Indentation fracture toughness of brittle materials for Palmqvist cracks, in *Fracture Mechanics of Ceramics*, vol. **5** (eds R.C. Bradt and M. Sakai), Plenum, New York, pp. 97–105.
15. Niihara, K. (1983) A fracture mechanics analysis of indentation-induced Palmqvist cracks in ceramics. *Journal of Materials Science Letters*, **2**, 221–223.
16. Miyoshi, T. (1985) A study on evaluation of K_{IC} for structural ceramics. *Transactions of the Japan Society of Mechanical Engineers, Series A*, **51**, 2489–2497.
17. Quinn, G.D. and Bradt, R.C. (2007) On the Vickers indentation fracture toughness test. *Journal of the American Ceramic Society*, **90**, 673–680.
18. Fischer, H. and Marx, R. (2002) Fracture toughness of dental ceramics: Comparison of bending and indentation method. *Dental Materials*, **18**, 12–19.
19. Maensiri, S., Laokul, P., Klinkaewnarong, J. and Amornkitbamrung, V. (2007) Carbon-nanofiber-reinforced alumina nanocomposites: Fabrication and mechanical properties. *Materials Science and Engineering A*, **447**, 44–50.
20. Liu, T.X., Phang, I.Y., Shen, L. and Liu, T.X. (2004) Carbon nanotubes reinforced nylon 6 composite prepared by simple melt-compounding. *Macromolecules*, **37**, 7214–7222.
21. Wang, X., Padture, N.P. and Tanaka, H. (2004) Contact-damage-resistant ceramic/single-wall carbon nanotubes and ceramic/graphite composites. *Nature Materials*, **3**, 539–544.
22. Padture, N.P. and Curtin, W.A. (2008) Comment on "Effect of sintering temperature on a single-wall carbon nanotube toughened alumina-based nanocomposite". *Scripta Materialia*, **58**, 989–990.
23. Jiang, D. and Mukherjee, A.K. (2008) Response to comment on "Effect of sintering temperature on single-wall carbon nanotube toughened alumina-based nanocomposite". *Scripta Materialia*, **58**, 991–993.
24. Chen, Y., Balani, K. and Agarwal, A. (2008) Analytical model to evaluate interface characteristics of carbon nanotube reinforced aluminum oxide nanocomposites. *Applied Physics Letters*, **92**, 0119161–0119163.
25. Guo, S., Sivakumar, R. and Kagawa, Y. (2007) Multiwall carbon nanotube-SiO_2 nanocomposites: Sintering, elastic

properties and fracture toughness. *Advanced Engineering Materials*, **9**, 84–87.

26 Veprek, S. (1999) The search for novel, superhard materials. *Journal of Vacuum Science & Technology A-Vacuum Surfaces and Films*, **17**, 2401–2420.

27 Voevodin, A.A. and Zabinski, J.S. (2005) Nanocomposite and nanostructured tribological materials for space applications. *Composites Science and Technology*, **65**, 741–748.

28 Lim, D.S., You, D.H., Choi, H.J., Lim, S.H. and Jang, H. (2005) Effect of CNT distribution on tribological behavior of alumina-CNT composites. *Wear*, **259**, 539–544.

29 An, J.W., You, H. and Lim, D.S. (2003) Tribological properties of hot-pressed alumina-CNT composites. *Wear*, **255**, 677–681.

30 Xia, Z.H., Lou, J. and Curtin, W.A. (2008) A multiscale experiment on the tribological behavior of aligned carbon nanotube/ceramic composites. *Scripta Materialia*, **58**, 223–226.

8
Conclusions

8.1
Future Prospects

Research and development efforts throughout the academic and industrial sectors have resulted in innovative technologies for the fabrication of carbon nanotubes (CNTs) with reduced weight, and excellent electrical, mechanical and thermal characteristics. Consequently, CNTs offer tremendous opportunities for the development of advanced functional materials. Numerous researchers have reported remarkable improvement in the mechanical properties of metals and ceramics by adding low loading levels of CNTs. Carbon nanotubes can strengthen metals and ceramics through the load-transfer effect, and toughen brittle ceramics via a crack-bridging mechanism. Composites reinforced with CNTs also exhibit excellent electrical and thermal conductivities. These composites show potential applications for thermal management in electronic devices, particularly in the telecommunications sector. However, heat dissipation is a key issue that limits the performance and reliability of electronic devices from cellular phones to satellites. Such commercial products require miniature composite materials with multifunctional properties. Furthermore, CNTs show remarkable biocompatibility and bioactivity, rendering them attractive materials for clinical applications. Generally, fundamental understanding of the synthetic process for achieving homogeneous dispersion of nanotubes and the structure–property relationship of nanocomposites is still lacking. From a more fundamental perspective, the principles of reinforcement, deformation, mechanical failure, electrical and heat transport of these nanocomposites are not completely understood and tested. Proper understanding of the processing– structure–property relationship is critical for designing novel nanocomposites with functional properties for specific applications. Up till now, only a few inventions have been acknowledged by global patent authorities for the possible commercialization of CNT-reinforced composites (Table 8.1) [Chap. 2, Ref. 99, Chap. 7, Ref. 3, 1–4]. This is due to the low production yield and high cost of CNTs. Obviously there is much to be done in enhancing the production yield of high quality nanotubes through technological innovations. Further commercial development of CNT-reinforced composites depends greatly on the availability of nanotubes at reasonable prices. For successful

Table 8.1 Patent processes for making ceramic- and metal-matrix nanocomposites reinforced with CNTs.

Inventors	Patent number	Patent year	Types of matrix materials	Fabrication Process	Description
Chang et al. Ref [1]	US 6420293	2002	Nanocrystalline ceramic oxides, nitrides, carbides, carbonitrides, and so on	Powder mixing	Powder blending followed by hot pressing
Zhan et al. Ref [2]	US 6858173	2005	Nanocrystalline alumina	Powder mixing	Mixing alumina nanoparticles with SWNTs followed by SPS
Zhan et al. Ref [Chap. 6, Ref. 3]	US 6875374	2005	Nanocrystalline alumina	Powder mixing	Mixing alumina nanoparticles with SWNTs followed by SPS
Zhan et al. Ref [Chap. 5, Ref. 40]	US 6976532	2005	Nanocrystalline alumina	As above	As above
Mamoru et al. Ref [3]	JP2006282489	2006	Hydroxyapatite	Powder mixing	Powder blending and SPS
Barrera et al. Ref [4]	US 7306828	2007	Ceramics	Powder mixing	Dry milling ceramic particles with CNTs, adding solvent to form a slurry, shape-forming the slurry into a green body, and sintering
Hong et al. Ref [Chap. 2, Ref. 99]	US 7217311	2007	Metals	In situ chemical precipitation	Dispersion of CNTs in a solvent, followed by mixing with metal salts ultrsonically, drying of the mixture, calcination and reduction in H_2, CO or CO_2 atmosphere

commercialization of CNT-reinforced composites, the fabrication and consolidation methods must be both cost effective and competitive.

Nanomaterials with unique physical properties, such as large surface area or aspect ratio have shown great potential for biological applications, including tissue engineering, biosensors, medical devices, and so on. The application of nanotechnology to biomedical engineering is a new frontier in orthopedics. Nanocomposites play a crucial role in orthopedic research since bone itself is a typical example of a nanocomposite [5, 6]. The fabrication of implant materials that mimic the structure and properties of human bones is a significant challenge for materials scientists. Bone is a composite material consisting of a calcium phosphate crystalline phase and an organic collagen matrix. The calcium phosphate phase in bones differs in composition from stoichiometric hydroxyapatite $[Ca_{10}(PO_4)_6(OH)_2]$ by the presence of other ions such carbonate, magnesium and fluoride of which the carbonate content is about 4–8 wt% [7]. Sintered hydroxyapatite (HA) cannot be used as a stand-alone load-bearing material for bones due to its brittle nature, so HA is mostly used as a coating material for bulk metallic implants. Therefore, CNTs with large aspect ratio, excellent flexibility, superior biocompatibility and bone cell adhesion appear to be suitable reinforcements for HA and other bioceramics. Apparently, orthopedic implant research is another area in which CNTs and their nanocomposites can be of significant importance [8]. Recently, Wang *et al.* reported that spark plasma sintered SiC/MWNT nanocomposites exhibit superior mechanical performance and biocompatibility [9]. These nanocomposites can be used for bone tissue repair and dental implantation on the basis of *in vivo* animal (rat) tests. Histological examination showed that there was little inflammatory response in the subcutaneous tissue, and newly formed bone tissue was observed in the femur after implantation for four weeks [9].

8.2
Potential Applications of CNT–Ceramic Nanocomposites

In addition to thermal management applications in the electronic industries, ceramic–CNT nanocomposites with good thermal conductivity and fracture toughness are candidates for leading edge applications such as thermal barrier coatings (TBCs) for gas turbine engines since they can improve the performance of turbine blades operate at extreme thermal and mechanical conditions [4]. Ceramics with reduced thermal conductivity have been used as thermal barrier materials for TBCs of gas turbine engines. The CNT–ceramic nanocomposites for TBC applications can be fabricated by means of tape-casting or plasma-spraying techniques. Another possible area where ceramic–CNT nanocomposites can find useful application is light-weight armor made from boron carbide [10]. The inherently brittle nature of boron carbide can be overcome by incorporating CNTs, thus improving the ballistic performance considerably.

It is anticipated that this novel emerging technology, nanotechnology, will have a substantial impact on biomedical engineering technology. At present, the development of ceramic–CNT nanocomposites for biomedical engineering applications is still in its infancy. Such nanocomposites should possess excellent biocompatibility and mechanical properties similar to those of natural bones. The success or failure of orthopedic implants depends on the cell–surface behavior after embedding into the human body. In this regard, the potential application of HA/CNT nanocomposites as future hard tissue replacement implants in the biomedical engineering sector is considered in the next section.

8.2.1
Hydroxyapatite–CNT Nanocomposites

Austenitic stainless steel, metallic Co–Cr–Mo and titanium-based alloys are widely used as implanted materials for artificial hip prostheses. They are suitable for load-bearing applications due to their combination of mechanical strength and toughness. However, the elastic modulus of metallic implants is not well matched with that of human bone, resulting in a stress shielding effect that can lead to reduced stimulation of bone tissue adhesion and growth. Furthermore, metallic alloys often suffer from pitting corrosion and stress-corrosion cracking upon exposure to human body environment. In this respect, ceramic composite materials offer distinct advantages over metallic alloys as implanted materials. It is recognized that ceramics possess excellent biocompatibility with bone cells and tissues. The intrinsic brittleness of ceramics precludes their use as bulk implants for load-bearing functions. Therefore, CNT additions are needed in order to improve the toughness of ceramics. Moreover, CNT incorporation is beneficial in enhancing the wear resistance of ceramics. Wear failure of conventional implants that results in joint loosening is a potential threat to rehabilitation of the patients. Failed implants require additional surgical operations that markedly increase cost and recovery time. Successful performance of the HA–CNT nanocomposites depends greatly on the nanotube content and fabrication process.

Hydroxyapatite is the main mineral constituent of human bones and teeth. An HA coating is generally deposited onto metal implants via plasma spraying. The major drawback of HA coating is its long term instability; HA tends to decompose into tricalcium phosphate (TCP), tetracalcium phosphate (TTCP) and non-biocompatible CaO during plasma spraying at elevated temperatures [11]. Delamination of HA coating from metal implants occurs readily due to its weak bond strength and chemical stability [12–14].

Bulk HA compacts prepared by conventional sintering generally exhibit low yield strength and fracture toughness [15]. High sintering temperature and long sintering time often result in grain coarsening and decomposition of HA. The fracture toughness of HA does not exceed $1\,\mathrm{MPa\,m^{1/2}}$ and much lower compared with that of human bone (2–$12\,\mathrm{MPa\,m^{1/2}}$) [16]. In this respect, spark plasma sintering (SPS) with relatively low temperature and short sinter duration can be used to consolidate HA to obtain products with better mechanical property and reduced grain size [17]. Recently, there has been an increasing interest in the processing of HA-based ceramics with nanometer grain sizes [18]. This is because the features of synthetic

HA nanocrystals resemble more closely the mineral constituents of human bone. Moreover, HA nanocrystals facilitate osteoblast (bone-forming cell) adhesion and cell proliferation [19, 20]. The molecular building blocks of cells such as proteins, nucleic acids, lipids and carbohydrates are confined to nanometer scales.

In this context, nano-HA with a large surface area enhances cell activities and promotes adsorption of proteins from body fluids [21–23]. In general, grain refinement of HA to nanoscale enhances its strength, but reduces its toughness considerably. To restore its toughness, CNTs with excellent high flexibility are incorporated into nano-HA. It has been reported that CNTs exhibit excellent biocompatibility and superior osteoblast adhesion [24, 25]. In this respect, biocomposites having nano-nano structure are considered to be potent implant materials for bone replacement function. Up till now, very little information is available in the literature relating the synthesis and structure of nano-HA/CNT composites [26, 27].

Zhao and Gao used *in situ* chemical precipitation to prepare nano-HA/MWNT nanocomposite powders [27]. In the process, MWNTs were dispersed initially in anionic SDS solution, forming electronegative charges on nanotube surfaces [Chap. 5, Ref. 62]. Carbon nanotube suspension was then added to calcium nitrate solution in which Ca^{2+} ions adsorb preferentially onto MWNTs due to the electrostatic attraction. Subsequently, di-ammonium hydrogen phosphate [$(NH_4)_2HPO_4$] was dissolved in dilute $NH_3 \cdot H_2O$ solution and added to the nanotube mixture solution. The PO_4^{3-} ions reacted *in situ* with Ca^{2+}, forming amorphous HA precipitates on MWNTs (Figure 8.1(a)). Precipitation of nano-HA on nanotubes occurred via the following reaction [28]:

$$10Ca(NO_3)_2 + 6(NH_4)_2HPO_4 + 8NH_4OH \rightarrow Ca_{10}(PO_4)_6(OH)_2 + 20NH_4NO_3 + 6H_2O.$$

Nanocrystalline HA precipitates can be obtained through proper hydrothermal treatment (Figure 8.1(b) and Figure 8.2). Such nano-HA/2 wt% MWNT powders were hot-pressed at 1200 °C. The compressive strength of nano-HA/2 wt% MWNT

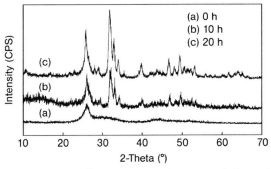

Figure 8.1 X-ray diffraction patterns of HA/MWNT before and after hydrothermal treatments for 10 and 20 h. Reproduced with permission from [27]. Copyright © (2004) Elsevier.

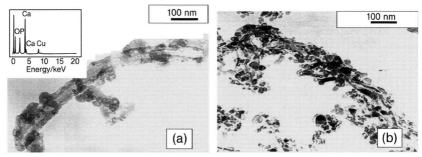

Figure 8.2 TEM micrographs of HA/MWNT (a) before and (b) after hydrothermal treatment for 20 h. Inset in (a) is EDX spectrum of HA/MWNT. Copper in the spectrum originates from sample holder. Reproduced with permission from [27]. Copyright © (2004) Elsevier.

composite is 102 MPa, compared with 63 MPa of pure HA. This corresponds to about 63% improvement in mechanical strength.

In the case of micro-nano HA/CNT composites, several processing techniques such as laser surface alloying [29, 30], plasma spraying [31, 32], sintering and hot pressing [33–35] and SPS [3] have been used to fabricate them. The fabrication processes employed for the fabrication of HA–CNT nanocomposites are summarized in Table 8.2. The high temperature environments of laser surface alloying and plasma spraying can cause serious degradation of the quality of CNTs. Further, laser surface alloying route results in the formation of undesired TiC phase [29]. Thus, pressureless sintering at lower temperature (1100 °C) [33] and SPS with short processing time [3] are more effective processes to consolidate the HA and CNT powder mixtures.

The HA/CNT powder mixtures can be prepared by the *in situ* synthesis route or through proper mixing of HA and nanotubes. The *in situ* route can be further classified into chemical vapor deposition (CVD) synthesis [36] and solution precipitation techniques [37, 38]. The *in situ* CVD technique involves the initial formation of Fe_2O_3/HA precursor, followed by calcination in a nitrogen atmosphere and reduction in hydrogen of the Fe_2O_3/HA precursor to yield Fe/HA catalyst and a final exposure to CH_4/N_2 gas mixture at 600 °C [38]. Figure 8.3(a) and (b) show typical SEM and TEM images of the *in situ* synthesized Fe-HA/CNT powder. As expected, the length of CVD-grown nanotubes is of the order of several tens of micrometers (Figure 8.3(a)). HA particles are strongly bonded and accumulated on the outer surface of MWNTs (Figure 8.3(b)). The synthesized HA/CNT powders were cold compacted and sintered at 1000 °C in vacuum for 2 h. As aforementioned, hot pressing *in situ* CVD-synthesized Fe-Al_2O_3/CNT nanocomposite powder at 1500 °C yields low density and poor mechanical strength due to the degradation of CNTs at high temperatures [Chap. 5, Ref. 42, Chap. 5, Ref. 59]. The synthesized Fe-HA/CNT powders were cold compacted and sintered at 1000 °C, thus the integrity of CNTs is preserved. As a result, sintered Fe-HA/CNT nanocomposite containing 2 wt% CNT and 1.5 wt% Fe exhibits a 226%

Table 8.2 Synthesis of micro-HA/CNT and nano-HA/CNT composite materials from different techniques.

Type of materials	Dispersion route	Heat treatment and/ or consolidation	Mechanical performance
Composite coating Ref [31]	Dry powder blending	Plasma spraying	Fracture toughness enhancement by 56%
Composite coating Ref [30]	Ball milling	Laser surface alloying	Hardness and elastic modulus enhancement
Bulk Ref [33]	Wet powder mixing	Cold isostatic pressing & pressureless sintering	Fracture toughness enhancement by more than 200%
Bulk Ref [3]	Wet powder mixing	Spark plasma sintering	—
Bulk Ref [36]	In situ CVD synthesis	Sintering	Fracture toughness and flexural strength enhancement
HA coating on MWNTs Ref [37]	In situ chemical precipitation	Vacuum drying at room temperature	—
HA coating on MWNTs Ref [38]	In situ chemical precipitation	Air drying at room temperature	—
Bulk nano-HA/CNT Ref [27]	In situ chemical precipitation	Hot pressing	Compressive strength enhancement

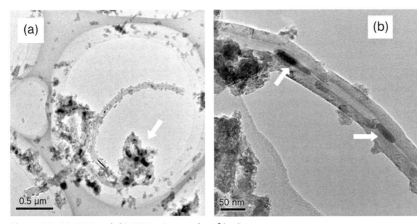

Figure 8.3 (a) SEM and (b) TEM micrographs of *in situ* synthesized HA/MWNT powders by CVD method. The arrows indicate Fe nanoparticles. Reproduced with permission from [36]. Copyright © (2008) Elsevier.

increase in Vickers indentation fracture toughness (2.35 MPa m$^{1/2}$) and 49% increase in flexural strength (79 MPa) compared with bulk monolithic HA having toughness of 0.72 MPa m$^{1/2}$ and strength of 53 MPa. Despite the fact that the *in situ* CVD technique produces an improvement in mechanical properties, the use of transition metal catalysts (e.g. Ni, Fe, Co, etc.) for CNT synthesis could have a negative effect on human health, particularly use of nickel. As reported, CNTs have excellent biocompatibility, but they may cause toxicity to human skin and lungs. The risk of human organ exposure to CNTs during nanotube synthesis and composite fabrication is ever increasing [43, 44]. Therefore, extensive biocompatibility and toxicological assessments of HA/CNT nanocomposites must be performed prior to their implantation to human body.

Lie *et al.* used prepared HA/3 wt%MWNT nanocomposites by wet powder mixing method followed by cold isostatic pressing and pressureless sintering at 1100 °C in different environments (air, argon and vacuum) for 3 h [33]. They reported that sintering the powder compacts of HA and MWNT in vacuum environment yields higher bending strength and fracture toughness than those sintered in air or argon. This is because sintering in vacuum eliminates pores in the nanocomposite specimen. The bending strength and fracture toughness of vacuum and argon sintered HA/3 wt%MWNT nanocomposite are 66.11 MPa and 2.40 MPa m$^{1/2}$ and 61.43 MPa, and 0.761 MPa m$^{1/2}$, respectively. The nanocomposite specimens were then implanted into the muscle of big white rats for biocompatibility testing. For the purpose of comparison, ZrO$_2$/HA composite samples were also implanted into the muscle of rats. Because of its high strength and stress-induced phase transformation toughening, zirconia has been used to strengthen HA. The ZrO$_2$/HA composite has been used as a coating material for biomedical implants [39, 40]. Figure 8.4(a) and (b) and Figure 8.5(a) and (b) show representative pathology micrographs of the HA/MWNT and ZrO$_2$/HA composites after short term implantation into muscle of rats. It is apparent that the inflammatory cell response in tissue is more serious for the ZrO$_2$/HA composite after implantation for one and three days. This implies that the

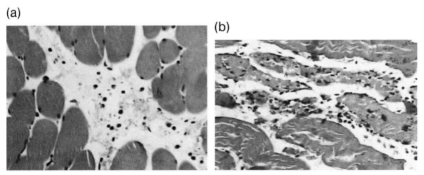

Figure 8.4 Tissue responses of (1) HA/MWNT and (b) ZrO$_2$/HA composite specimens after implantation into muscle of rats for one day. Reproduced with permission from [33]. Copyright © (2007) Elsevier.

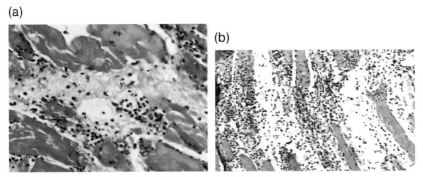

Figure 8.5 Tissue responses of (1) HA/MWNT and (b) ZrO$_2$/HA composite specimens after implantation into muscle of rats for three days. Reproduced with permission from [33]. Copyright © (2007) Elsevier.

HA/MWNT nanocomposite exhibits better biocompatibility than ZrO$_2$/HA composite. The outcome of *in vivo* experiment for HA/MWNT nanocomposite is similar to that of SiC/MWNT nanocomposites [9]. More studies are needed in near future to improve the fabrication process, mechanical property and biocompatibility of HA/MWNT nanocomposites for biomedical applications.

8.3
Potential Applications of CNT–Metal Nanocomposites

Conventional metal-matrix composites reinforced with ceramic materials and carbon fibers found extensive structural applications in aerospace, automotive and transportation industries. The incorporation of ceramic reinforcements and carbon fibers into metal matrices increases the tensile strength and stiffness but degrades the ductility markedly. Toughness is a key factor that influences the performance of metal-matrix composites for various engineering applications. Carbon fibers produced from PAN precursor have reached their performance limit. Carbon nanotubes with superior flexibility can overcome the inherent problems of ceramic fillers and carbon fibers. Thus, multifunctional composites with improved mechanical, electrical and thermal properties can be prepared by adding low level content of nanotubes. Such nanocomposites are considered to be an important new class of structural materials for the mechanical components of microelectromechanical systems such as high-frequency micromechanical resonator devices [42]. The main obstacles for commercialization of CNT-reinforced metals are the high cost of CNTs and agglomeration of fillers in metal matrices. There will certainly be more effective applications of light-weight CNT-reinforced metals as structural and functional materials in the foreseeable future through novel improvement in processing techniques and cost reduction.

References

1 Chang, S., Doremus, R.H., Siegel, R.W. and Ajayan, P.M. (2002) Ceramic nanocomposites containing carbon nanotubes for enhanced mechanical behavior. US Patent 6420293.

2 Zhan, G., Mukherjee, A.K., Kuntz, J.D. and Wan, J. (2005) Nanocrystalline ceramic materials reinforced with single-wall carbon nanotubes. US Patent 6858173.

3 Mamoru, O. and Toshiyuki, H. (2006) Composite material composed of carbon nanotube and hydroxyapatite, and method for producing the same. Japan Patent JP2006282489.

4 Barrera, E.V., Yowell, L.L., Mayeaux, B.M., Corral, E.L. and Cesarano, J. (2007) Fabrication of reinforced composite material comprising carbon nanotubes, fullerenes, and vapor-grown carbon fibers for thermal barrier materials, structural ceramics, and multifunctional nanocomposite ceramics. US Patent 7306828.

5 Stevens, M.M. and George, J.H. (2005) Exploring and engineering the cell surface interface. *Science*, **310**, 1135–1138.

6 Christenson, E.M., Anseth, K.S., van den Beucken, J.J., Chan, C.K., Ercan, B., Jansen, J.A., Laurencin, C.T., Li, W.J., Murugan, R., Nair, L.S., Ramakrishna, S., Tuan, R.S., Webster, T.J. and Mikos, A.G. (2007) Nanobiomaterial applications in orthopedics. *Journal of Orthopaedic Research*, **25**, 11–22.

7 Driessens, F.C.M. (1983) *Bioceramics of Calcium Phosphates*, CRC Press, pp. 1–32.

8 Harrison, B.S. and Atala, A. (2007) Carbon nanotube applications for tissue engineering. *Biomaterials*, **28**, 344–353.

9 Wang, W., Watari, F., Omori, M., Liao, S., Zhu, Y., Yokoyama, A., Uo, M., Kimura, H. and Ohkubo, A. (2007) Mechanical properties and biological behavior of carbon nanotube/polycarbosilane composites for implant materials. *Journal of Biomedical Materials Research Part B: Applied Biomaterials*, **82**, 223–230.

10 Palicka, R.J. and Negrych, J.A. (1989) Method of manufacturing boron carbide armor tiles. US Patent 4824624.

11 Sun, L., Berndt, C.C. and Grey, C.P. (2003) Phase, structural and microstructural investigations of plasma sprayed hydroxyapatite coatings. *Materials Science and Engineering A*, **360**, 70–84.

12 Cheang, P. and Khor, P. (1996) Addressing problems associated with plasma spraying of hydroxyapatite coatings. *Biomaterials*, **17**, 537–544.

13 Bauer, T.W., Geesink, R.C., Zimmerman, R. and McMahon, J.T. (1991) Hydroxyapatite-coated femoral stems. Histological analysis of components retrieved at autopsy. *Journal of Bone and Joint Surgery-American Volume*, **73**, 1439–1452.

14 Kueh, S.W., Khor, K.A. and Cheang, P. (2002) An *in vitro* investigation of plasma sprayed hydroxyapatite (HA) coatings produced with flame-spheroidized feedstock. *Biomaterials*, **23**, 775–785.

15 Slosarczyk, A. and Bialoskorski, J. (1998) Hardness and fracture toughness of dense calcium-phosphate-based materials. *Journal of Materials Science-Materials in Medicine*, **9**, 103–108.

16 Suchanek, W. and Yoshimura, M. (1998) Processing and properties of hydroxyapatite-based biomaterials for use as hard tissue replacement implants. *Journal of Materials Research*, **13**, 94–117.

17 Gu, Y.W., Loh, N.H., Khor, K.A., Tor, S.B. and Cheang, P. (2002) Spark plasma sintering of hydroxyapatite powders. *Biomaterials*, **23**, 37–43.

18 Norton, J., Malik, K.R., Darr, J.A. and Rehman, I, (2006) Recent developments in processing and surface modification of hydroxyapatite. *Advances in Applied Ceramics*, **105**, 113–139.

19 Webster, T.J., Ergun, C., Doremus, R.H., Siegel, R.W. and Bizios, R. (2000) Enhanced functions of osteoblasts on nanophase ceramics. *Biomaterials*, **22**, 1803–1810.

20 Adamopoulos, O. and Papadopoulos, T. (2007) Nanostructured bioceramics for maxillofacial applications. *Journal of Materials Science-Materials in Medicine*, **18**, 1587–1597.

21 Cai, Y., Liu, Y., Yan, W., Hu, Q., Tao, J., Zhang, M., Shi, Z. and Tang, R. (2007) Role of hydroxyapatite nanoparticles size in bone proliferation. *Journal of Materials Chemistry*, **17**, 3780–3787.

22 Webster, T.J., Siegel, R.W. and Bizios, R. (1999) Osteoblast adhesion on nanophase ceramics. *Biomaterials*, **20**, 1221–1227.

23 Webster, T.J., Schadler, L.S., Siegel, R.W. and Bizios, R. (2001) Mechanisms of enhanced osteblast adhesion on nanophase alumina involve vitronectin. *Tissue Engineering*, **7**, 291–301.

24 Lin, Y., Taylor, S., Li, H., Fernando, K.A., Qu, L., Wang, W., Gu, L., Zhou, B. and Sun, Y.P. (2004) Advances toward bioapplications of carbon nanotubes. *Journal of Materials Chemistry*, **14**, 527–541.

25 Smart, S.K., Cassady, A.I., Lu, G. and Martin, D.J. (2006) The biocompatibility of carbon nanotubes. *Carbon*, **44**, 1034–1047.

26 Liao, S., Xu, G., Wang, W., Watari, F., Cui, F., Ramakrishna, S. and Chan, C.K. (2007) Self-assembly of nano-hydroxyapatite on multi-walled carbon nanotubes. *Acta Biomaterialia*, **3**, 669–675.

27 Zhao, L. and Gao, L. (2004) Novel *in situ* synthesis of MWNTs-hydroxyapatite composites. *Carbon*, **42**, 423–426.

28 Jie, W., Bao, L.Y. and Yi, H. (2005) Processing properties of nano apatite-polyamide biocomposites. *Journal of Materials Science*, **40**, 793–796.

29 Chen, Y., Gan, C., Zhang, T. and Yu, G. (2005) Laser-surfaced-alloyed carbon nanotubes reinforced hydroxyapatite composite coatings. *Applied Physics Letters*, **86**, 251905.

30 Chen, Y., Zhang, T., Zhang, T.H., Gan, C.H., Zheng, C.Y. and Yu, G. (2006) Carbon nanotube reinforced hydroxyapatite composite coatings produced through laser surface alloying. *Carbon*, **44**, 37–45.

31 Balani, K., Anderson, R., Laha, T., Andara, M., Tercero, J., Crumpler, E. and Agarwal, A. (2007) Plasma-sprayed carbon nanotube reinforced hydroxyapatite coatings and their interaction with human osteoblasts *in vitro*. *Biomaterials*, **28**, 618–624.

32 Balani, K., Chen, Y., Harimkar, S.P., Dahotre, N.B. and Agarwal, A. (2007) Tribological behavior of plasma-sprayed carbon nanotube-reinforced hydroxyapatite coating in physiological solution. *Acta Biomaterialia*, **3**, 944–951.

33 Li, A., Sun, K., Dong, W. and Zhao, D. (2007) Mechanical properties, microstructure and histocompatibility of MWNTs/HAp. *Materials Letters*, **61**, 1839–1844.

34 Kobayashi, S. and Kawai, W. (2007) Development of carbon nanofiber reinforced hydroxyapatite, with enhanced mechanical properties. *Composites A*, **38**, 114–123.

35 Kealley, C., Elcombe, M., Riessen, A. and Ben-Nissan, B. (2006) Development of carbon nanotube-reinforced hydroxyapatite bioceramics. *Physica B*, **385–386**, 496–498.

36 Li, H., Zhao, N., Liu, Y., Liang, C., Shi, C., Du, X. and Li, J. (2008) Fabrication and properties of carbon nanotubes reinforced, Fe/hydroxyapatite composites by *in situ* chemical vapor deposition. *Composites A*, **39**, 1128–1132.

37 Zhao, B., Hu, H., Mandal, S.K. and Haddon, R.C. (2005) A bone mimic based on the self-assembly of hydroxyapatite on chemically functionalized single-walled carbon nanotubes. *Chemistry of Materials*, **17**, 3235–3241.

38 Aryal, S., Remant Bahadur, K.C., Dharmaraj, N., Kim, K.W. and Kim, H.Y. (2006) Synthesis and characterization of hydroxyapatite using carbon nanotube as a

nano-matrix. *Scripta Materialia*, **54**, 131–135.

39 Yoshida, K., Hashimoto, K., Toda, Y., Udagawa, S. and Kanazawa, T. (2006) Fabrication of structure-controlled hydroxyapatite/zirconia composite. *Journal of the European Ceramic Society*, **26**, 515–518.

40 Lee, T.M., Tsai, T.S., Chang, E., Yang, C.Y. and Yang, M.R. (2002) Biological responses of neonatal rat calvarial osteoblasts on plasma sprayed HA/ZrO_2 composite coating. *Journal of Materials Science-Materials in Medicine*, **13**, 281–287.

41 Lee, T.M., Yang, C.Y., Chang, E. and Tsai, R.S. (2004) Comparison of plasma-sprayed hydroxyapatite coatings and zirconia-reinforced hydroxyapatite composite coatings: *in vivo* study. *Journal of Biomedical Materials Research Applied Biomaterials*, **71**, 652–660.

42 Bak, J.H., Kim, Y.D., Hong, S.S., Lee, B.Y., Lee, S.R., Jang, J.H., Kim, M., Char, K., Hong, S. and Park, Y.D. (2008) High-frequency micromechanical resonators from aluminum-carbon nanaotube nanolaminates. *Nature Materials*, **7**, 459–463.

43 Grabinski, C., Hussain, S., Lafdi, K., Braydich-Stolle, L. and Schlager, J. (2007) Effect of particle dimension on biocompatibility of carbon, nanomaterials. *Carbon*, **45**, 2828–2835.

44 Tian, F., Cui, D., Schwarz, H., Estrada, G.G. and Kobayashi, H. (2006) Cytotoxicity of single- wall carbon nanotubes on human fibroblasts. *Toxicology in Vitro*, **20**, 1202–1212.

Index

a
agglomerates 20
alumina ceramics 138
aluminum 47
aspect ratio 1
atomic force microscopy 26

b
ball milling 20
bath composition 73
bending 25
biocompatibility 217

c
carbon nanotubes 1
carbon nanotubes
– Raman spectroscopy 21
– synthesis of 5
carbonaceous species 17
carboxyl (-COOH) groups 19
catalyst 5
ceramic particulates 44
chemical vapor deposition 8
chirality 4
clusters 46
coating 48
CO disproportionation 16
colloidal processing 140
conductive network 171
consolidation 52
copper 65
crack bridging 133
crack deflection 133

d
deformation 25
dielectric constant 172
disintegrated melt deposition 61
ductility 57

e
electrical conductivity 171
electric arc discharge 5
electrodeposition 71
electromagnetic interference (EMI) 169
elongation 29
extrusion 45

f
fiber 1
flexibility 46
fracture toughness 1
friction, coefficient of 119
functional groups
– carboxyl (-COOH) groups 19

g
grain refinement 58
graphene 3
– multi-walled carbon nanotubes 3
– single-walled carbon nanotubes 3

h
hardness 105
heat dissipation 89
heterocoagulation 145
homogeneous dispersion 47
hot isostatic pressing 45
hot pressing 44
human bones 217
hydrocarbon gases 10
hydroxyapatite 217

i
interface
– interfacial bonding 47

l
laser ablation   7
load transfer   47, 103

m
magnesium   61
MD simulation   25
mechanical alloying   53
melt infiltration   45
metal-matrix composites   1
micromechanical model   103
microparticles   9
molecular-level mixing   47

n
nanocomposites   2
nanoindentation   105

o
Orowan stress   117

p
patent   17
percolation   169
powder metallurgy   44
pull-out of nanotubes   109
purification process   19

r
Raman spectroscopy   21
reinforcement   1
relative density   142

s
silicon carbide   157
silicon nitride   157
single edge notched beam   185
spark plasma sintering   67
stiffness   1
strain   25
strength   1
strengthening mechanism   103
stress intensity factor   185
SWNT, tensile elongation   29
synthesis   5

t
tension   45
thermal conductivity   29
thermal expansion, coefficient of   91
thermal management   45
thermal spraying   47
three-point bending   105
transmission electron micrograph   3

v
van der Waals forces   3
vapor grown carbon nanofibers   4
Vickers indentation fracture test   186

w
wear   119
wetting   47

y
Young's modulus   25

z
zeta potential   141